Fire Safety Design for Tall Buildings

Fire Safety Design for Tall Buildings

Feng Fu

CRC Press is an imprint of the
Taylor & Francis Group, an **informa** business

First edition published 2021
by CRC Press
6000 Broken Sound Parkway NW, Suite 300, Boca Raton, FL 33487-2742

and by CRC Press
2 Park Square, Milton Park, Abingdon, Oxon, OX14 4RN

First issued in paperback 2022

Visit the Taylor & Francis Web site at
http://www.taylorandfrancis.com

and the CRC Press Web site at
http://www.crcpress.com

ISBN: 978-0-367-44452-5 (hbk)
ISBN: 978-0-367-69771-6 (pbk)
ISBN: 978-1-003-00981-8 (ebk)

DOI: 10.1201/9781003009818

Typeset in Sabon
by codeMantra

To my Daughter Quan

Contents

Preface

I started to work on fire safety design for tall buildings since I joined WSP Group in 2007. I have been working on more than ten different types of buildings for fire safety design. I was very lucky to be one of the key team members for structural fire design of the currently tallest building in Western Europe, the Shard. To design fire safety for a building with such a complex nature is one of the most challenging jobs I have ever done. One of the major difficulties is that no systematic approach for structural fire design was available for tall buildings. Therefore, working together with my colleagues, we developed a systematic structural fire analysis framework for tall buildings via 3D finite element analysis in Abaqus®. The framework provided a cost-effective fire resistance design for the client.

After joining the academia, I was invited by Institution of Structural Engineers to give lecture for practicing structural engineers on tall building design with one session to introduce fire safety design for tall buildings. While teaching, I noticed that most of the structural engineers lack fire safety design knowledge and are desperate to gain understanding of the relevant design and analysis methods. I have also been invited by Charted Institute of Architectural Technologists to give talks on fire safety design for tall buildings; in doing so, I found that most architectural engineers were also keen to gain the relevant knowledge. However, there are few textbooks available for fire safety design of tall buildings.

Therefore, this textbook is designed to help fire safety engineering practitioners such as structural engineers, architectural engineers, and students to fully understand the principles of fire safety design and the relevant design guidance, particularly for tall buildings; explain effective ways to model different fire scenarios and analyze thermal response of building elements in fire; and introduce a systematic fire safety design approach for tall buildings.

Another feature of this book is that it demonstrates the 3D modeling techniques for fire safety analysis through the examples which replicate the real fire incidents such as Twin Tower, World Trade Center 7, and Cardington fire test. This would be helpful to the engineers in understanding the effective way to analyze the structures for fire safety design.

Feng Fu

Acknowledgments

I would like to express my gratitude to Dassault Systems and/or its subsidiaries, ANSYS Inc. and/or its subsidiaries, and Autodesk Inc. to give me the permission to use the images of their product.

I also thank BSI Group in U.K. and National Institute of Standards and Technology, Technology Administration, U.S. Department of Commerce for allowing me to reproduce some of the images from their reports.

Some of the models, drawings, and charts used in this book are made by me and some are based on the work of my MSc and final-year students. I am very appreciative to my final-year and MSc students: Mr Shariq Naqvi, Mr Wing Sing Tsang, Mr Zubair Aziz, Mr Yu Zhang, and Mr Xiao Li. I also thank Mr Xuandong Chen for his help.

I am thankful to all reviewers who offered their comments. Special thanks to Tony Moore and Frazer Merritt from Taylor & Francis for their assistance in preparation of this book.

Thanks to my family, especially my father Changbin Fu, my mother Shuzhen Chen, and my wife Dr. Yan Tan for their support in finishing this book.

Author

Dr. Feng Fu, Ph.D., MBA, CEng, FIStructE, FASCE, FICE, FHEA, worked for several world-leading consultancy companies including WSP Group, where he was one of the key team members in structural fire design of the tallest building in Western Europe, the Shard. Currently, he serves for two building design standard committees of the American Society of Civil Engineers and also acts as an associate editor and editorial board member for three international journals. He has published more than 100 technical papers and three textbooks including *Structural Analysis and Design to Prevent Disproportionate Collapse* (CRC Press, 2016).

Chapter 1

Introduction

1.1 AIMS AND SCOPE

Across the world, there are many fire incidents happening in tall or multistorey buildings every year. It causes loss of life and damage to the properties. About 69% fires are caused by electricity. As a result, fire safety is one of the key tasks in tall building designs. Particularly, a large percentage of tall buildings are steel structures or partially made of steel frames which require more stringent fire safety design requirements. The recent disaster in Grenfell Tower (Fu, 2017) caused huge casualties. It has embarked increasing concerns from the public and building engineers in fire safety design for tall buildings.

Fire development and subsequent thermal response of the building depend upon numerous factors, invariably featuring a high degree of uncertainty. While permitted within performance-based frameworks and supported by design codes (EN 1991-1-2, 2002; EN1992-1-2, 2004; EN 1993-1-2, 2005; EN 1994-1-2, 2005), the appraisal of structural response in fire is challenging given the sources of uncertainty that exist. This is primarily due to the complexity caused by different fire scenarios which can possibly be formed when fire occurs. In addition, for a structure such as a tall building, the structural systems are much more complicated, which also brings extra difficulties in the structural fire analysis.

In the past two decades, an increasing number of tall buildings have been built worldwide. Advancements in structural engineering make possible the increase in height, size, and complexity of modern tall buildings. Particularly, in modern tall building design, the structural system of tall buildings becomes increasingly complicated. Figure 1.1 shows the structural system of a newly built tall building in Beijing with a height of 528 m. The lateral stability system comprises so-called Mega Frame system and Diagrid system. Its more complex structural system makes its fire safety design a challenging job. In order to effectively design fire safety for tall buildings, it is essential to understand the behavior of the buildings in fire.

In addition, a significant amount of new construction materials—elements such as new type of cladding systems and new construction

Figure 1.1 China Zun Tower in construction. (Photo taken by the author's father.)

techniques—have also been developed. The tall building of today is different from that of a decade ago with foreseen changes even greater in the immediate future. These advancements make fire safety design for tall buildings an even more challenging task for design engineers.

In the current design practice, detailed fire safety design guidelines has been developed across the world, such as Eurocode (EN 1991-1-2, 2002; EN1992-1-2, 2004; EN 1993-1-2, 2005; EN 1994-1-2, 2005). As a design engineer, it is imperative to guarantee that in the design process, sufficient measures for fire safety should be made. An engineer should also have the capacity to analyze the response of structures under different fire scenarios and subsequent fire protection measures using appropriate procedure and analysis software.

Therefore, this textbook is designed to help fire safety engineering practitioners such as structural engineers, architectural engineers, and students to fully understand the principles of fire safety design and the relevant design guidance, particularly for tall buildings; the effective way to model different fire scenarios and thermal response of building in fire; and introduction of a systematic fire safety design approach for tall buildings. Detailed demonstrations of 3D modeling techniques for fire safety analysis are also made. In addition, case studies based on various fire scenarios and different structural layouts of tall buildings are provided to demonstrate failure mechanisms of buildings in fire and effective design methods for fire safety.

1.2 MAIN FIRE SAFETY DESIGN ISSUES FOR TALL BUILDINGS

The main objective of the fire safety design for tall buildings is life safety of the occupants. Therefore, all the design processes are centered around life safety. Among them, compartmentation design, evacuation route design, and structural fire design are the three key design focuses. These three factors are affecting each other, for example, when designing the evacuation route, the time of evacuation is affected by the time of failure of structural members. The size and layout of the compartment also affect the evacuation route design. The integrity of the compartment is very important in containing the fire in its original place or delaying its spread. However, the integrity of the compartment is greatly affected by structural fire design. For example, the deformation of the compartment wall in fire reduces its capacity to maintain its integrity. Its deflection is controlled primarily by its supporting beams. Designing a beam with reduced deflection in fire will improve the whole integrity of the compartment. These design processes will be explained in detail in Chapters 3–5.

Fire scenario analysis and its corresponding structural fire analysis are both unique and complicated procedures, and they also require an engineer to have the ability to use modern commercial software for fire scenarios simulation or a finite element package to analyze the structural responses of a building in fire. Therefore, this book also features a detailed introduction to the use of fire analysis software such as OZONE and FDS®[1] as well as finite element programs such as Abaqus®,[2] ANSYS, and LS_DYNA OpenSees, ADINA.

1.3 STRUCTURE OF THE BOOK

Chapter 1 is the introduction of the book. It introduces the aims and scope, as well as the structure, of this book.

Chapter 2 introduces several fire incidents that happened in tall buildings, followed by the regulatory requirements from various codes across the world. At the end of this chapter, the basic principles for fire safety design of tall buildings will be discussed.

Chapter 3 introduces the fundamental knowledge of fire and fire safety design. The characteristics of fire and its development are introduced at the beginning. The key fire scenarios that affect the performance of the building members in fire—such as ventilation-controlled or fuel-controlled fire and long-cool, short-hot fire—will be explained. In addition, the fundamentals of heat transfer, a process of the heating up of structural members due to fire, will be introduced. The basic structural fire design principles will also be explained. In fire safety design, most of the codes specify the fire resistance of building elements. The relevant information will be provided in the latter part of this chapter followed by the introduction of fire protection methods.

Chapter 4 introduces the structural fire design in depth on the basis of Chapter 3. It introduces the structural fire design procedures for steel, concrete, and composite structural members based on Eurocodes and British Standards. Two key structural fire design methods are introduced: critical temperature method and moment capacity method. It also covers the design of post-tensioning slabs, connection, and beams with openings.

Chapter 5 provides a detailed design strategy for tall buildings. The prescriptive and performance-based fire design approaches are first introduced, followed by fire risk analysis. The deterministic and probabilistic approach to determine the worst-case fire scenarios is then introduced. A detailed demonstration of compartment design and evacuation route design for tall buildings is made followed by the description of other design issues such as firefighter access and fire protection requirement to façade. The fire alarm system, communication system, fire and smoke suppression system are also discussed. At the end of this chapter, case studies for two real construction projects, namely, Burj Khalifa and the Shard, are made.

Chapter 6 introduces various theoretical and numerical methods for fire analysis. It starts with the method to determinate the compartment fire including a detailed introduction of Zone model and CFD model. It is followed by the methods of solving thermal response of structural members such as heat transfer analysis and thermal–mechanical analysis. In addition, the probabilistic method for fire safety analysis will be covered. In the final part of this chapter, various numerical modeling software for fire analysis will be explained.

Chapter 7 discusses how to design a building to prevent fire-induced collapse. The collapse mechanism of a tall building in fire and methods for mitigating the collapse are introduced, all based on existing research and fire-induced collapse incidents.

Chapter 8 introduces new technologies developed for fire safety design, such as PAVA system, IOT, and smart building management system. Some pilot studies of using machine leaning in fire safety design will also be introduced in this chapter.

Chapter 9 introduces the post fire damage assessment methods. Different damage assessment techniques including destructive and nondestructive assessment methods for concrete and steel structures are introduced.

NOTES

1 FDS https://www.nist.gov/services-resources/software/fds-and-smokeview.
2 Abaqus is a registered trademark of Dassault Systemes S.E. and Its affiliates.

REFERENCES

EN 1991-1-2 (2002), Eurocode 1. Actions on Structures-Part 1-2: General actions. Actions on structures exposed to fire. Commission of the European communities.

EN1992-1-2 (2004), Eurocode 2. Design of concrete structures, Part 1-2: General rules. Structural fire design. Commission of the European communities.

EN 1993-1-2 (2005), Eurocode 3. Design of steel structures, Part 1-2: General rules. Structural fire design. Commission of the European communities.

EN 1994-1-2 (2005), Eurocode 4. Design of composite steel and concrete structures, Part 1-2: General rules. Structural fire design. Commission of the European communities.

Fu, F. (2017), Grenfell Tower disaster: How did the fire spread so quickly? *BBC Australia*.

Chapter 2

Regulatory requirements and basic fire safety design principles

2.1 INTRODUCTION

In this chapter, several fire incidents that happened in tall buildings will be first introduced followed by the regulatory requirements from various codes across the world. At the end of this chapter, the basic principles for fire safety design of tall buildings will be discussed.

2.2 FIRE INCIDENTS AND FIRE TESTS OF TALL BUILDINGS WORLDWIDE

Table 2.1 shows the fire incidents occurred each year in different countries during 2012–2015. It can be seen that there are a large number of fire incidents happening each year across the world.

Table 2.1 Fire incidents across the world from 2012 to 2015 in different countries

No.	*Country*	*Population in 1,000*	*2012*	*2013*	*2014*	*2015*
1	US	323,128	1,375,000	1,240,000	1,298,000	1,345,500
2	Bangladesh	154,331	17,504	17,912	17,830	17,488
3	Russia	146,270	162,900	152,959	150,437	145,900
4	Japan	128,130	44,101	48,095	43,741	39,111
5	Vietnam	93,000	1,900	2,540	2,375	2,451
6	Germany	82,218	-	-	175,354	192,078
7	France	66,628	306,871	281,908	270,900	300,667
8	Great Britain	63,786	272,800	192,700	212,500	191,647
9	Italy	61,000	241,232	196,196	189,375	234,675

(*Continued*)

Table 2.1 (Continued) Fire incidents across the world from 2012 to 2015 in different countries

No.	*Country*	*Population in 1,000*	*2012*	*2013*	*2014*	*2015*
10	Myanmar	51,486	1,219	1,673	1,629	-
11	Spain	47,079	142,500	135,000	128,000	137,000
12	Ukraine	42,673	71,443	61,144	68,879	79,640
13	Poland	38,454	183,888	125,425	145,237	184,847
14	Canada	35,544	45,005	37,194	36,445	-
15	Malaysia	31,800	29,874	33,640	54,540	40,865
16	Peru	30,741	11,329	11,264	9,430	9,473
17	Nepal	30,430	-	1,021	958	-
18	Taiwan	23,069	1,574	1,451	1,417	1,704
19	Romania	20,121	38,077	-	-	26,247
20	Kazakhstan	17,500	16,145	13,621	14,477	14,452
21	Netherlands	16,979	-	-	91,160	125,200
22	Greece	10,788	33,731	28,232	-	-
23	Belgium	10,700	21,369	21,228	-	-
24	Czech Republic	10,579	20,492	16,563	17,388	20,232
25	Sweden	9,851	22,657	25,392	-	22,785
26	Hungary	9,830	37,106	20,177	19,536	21,056
27	Jordan	9,700	23,961	25,644	20,795	32,488
28	Belarus	9,505	34,505	7,151	7,489	7,339
29	Austria	8,740	42,213	40,395	43,336	45,349
30	Switzerland	8,372	14,304	12,893	11,658	12,477

Source: Fi. (https://www.ctif.org/sites/default/files/2018-06/CTIF_Report23_World_Fire_Statistics_2018_vs_2_0.pdf)

Most of the fire will cause local damages to the buildings; however, some may even causes collapse the entire buildings. The two famous examples of fire-induced building collapse are Twin Towers and WTC7. They will be introduced in detail in this chapter.

2.2.1 Grenfell Tower

Grenfell Tower fire (Fu, 2017) caused a great tragedy with 71 deaths and over 70 injuries. Among the 129 flats, occupants of 23 flats died and 223 people escaped. There are a number of factors in the design of the 24-storey tower that may have contributed to the speed and scale of the blaze. The fire started in a kitchen at lower level of the tower, and then the flame propagated through the cladding to the upper level of the tower, causing the fire to spread to almost the entire building (Figure 2.1).

Figure 2.1 Fire incidents in Grenfell Tower. (This file is licensed under the Creative Commons Attribution 4.0 International license, https://commons.wikimedia.org/wiki/File:Grenfell_Tower_fire_(wider_view).jpg.)

2.2.1.1 *The new cladding system*

As shown in Figures 2.2 and 2.3, it was reported that a new cladding was added 1 year before the fire. From Figure 2.4, it can be seen that there are three layers in the cladding. The material used for the cladding was primarily aluminum, with an extra layer of insulation in between the aluminum layers.

British Research Establishment conducted a detailed investigation after the fire (British Research Establishment, 2017). Both the cladding panels and the infilled insulation of the new façade were tested under fire. It was found that they were not good in terms of fire resistance. What's more, aluminum has high conductivity, so the cladding itself could have heated up very quickly, failing to prevent the fire from spreading through the windows and up the exterior of the block from one storey to another.

2.2.1.2 *Compartment and evacuation route for Grenfell Tower*

Most of the current guidelines across the world contain detailed design requirements for fire safety. But at the time Grenfell Tower was built (in 1974),

Figure 2.2 Grenfell Tower before refurbishment. (This file is licensed under the Creative Commons Attribution-Share Alike 2.0 Generic license, Attribution: Robin Sones, https://upload.wikimedia.org/wikipedia/commons/b/b4/Grenfell_Tower%2C_London_in_2009.jpg.)

the rules and regulations were not as stringent as now; therefore, most of the old buildings did not conform to the latest guidelines for fire safety design. Hence, it is imperative to update them by making installation of sprinklers, fire alarms, and extra fire evacuation staircases mandatory.

2.2.1.2.1 Compartmentation

One of the key strategies in fire safety design is to correctly design fire compartments to keep the fire from spreading quickly. To contain the fire in a

Figure 2.3 New cladding burnt by fire. (This file is licensed under the Creative Commons Attribution 4.0 International license, https://upload.wikimedia.org/wikipedia/commons/9/9a/Grenfell_Tower_fire_morning.jpg.)

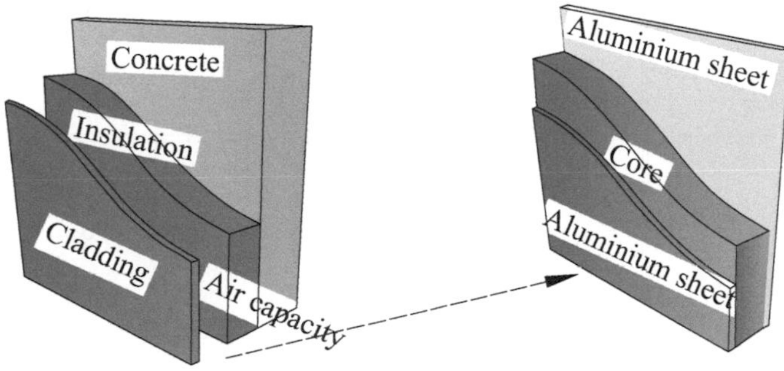

Figure 2.4 Schematic layout of the new cladding.

local area, placing barriers in the building—such as fire-resistant doors (fire doors) and walls (compartment walls)—is essential. These design measures can at least slow down the speed at which it spreads. These compartments are designed by architects based on the function of the buildings, so residential and commercial buildings will have different compartment design strategies.

The report of British Research Establishment (2017) has also noticed this. It is also reported that, ironically, fire doors were only installed in the storage rooms; therefore, all the stuff inside the storage room remained intact during the fire. No fire doors were installed in any flats. Figure 2.5 shows the structural layout of the Grenfell Tower, and no compartment wall and fire door were used apart from the fire door in the storage room in the entire floor, so the whole floor can be treated as one compartment, which means that when fire starts at one room, it will quickly spread into other rooms.

In the current design practice, some buildings even include special design measures for fires, such as refuge rooms in higher storeys for occupants who could have trouble escaping downstairs. There are also active fire protection methods such as using sprinklers. No sprinkler seems to be installed in Grenfell Tower. A local residents action group also claimed that their warnings about a lack of fire safety measures "fell on deaf ears."

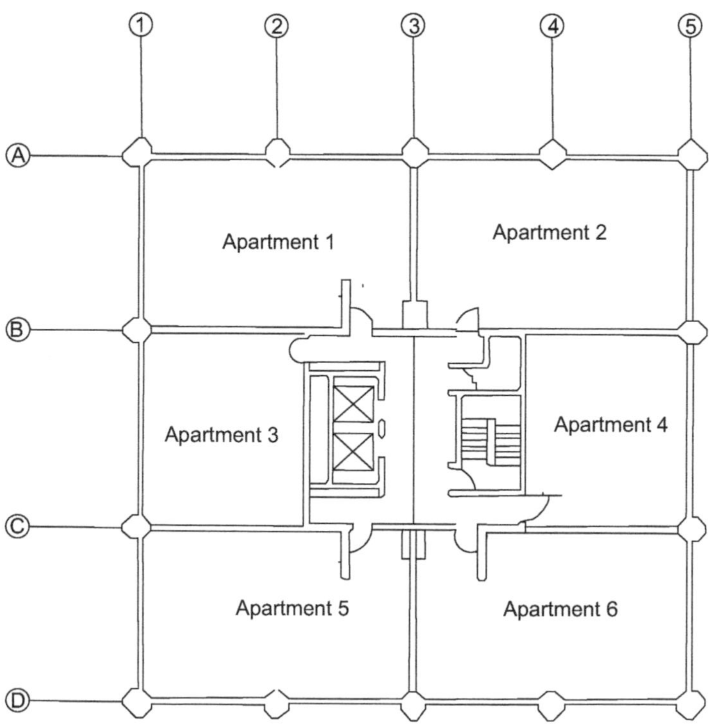

Figure 2.5 Compartment layout of Grenfell Tower.

2.2.1.2.2 Evacuation

The evacuation route is another most important design element in fire safety. For tall buildings, staircase is the major evacuation route for the occupants. The route should allow occupants to escape the building as quickly as possible while sheltering them from smoke and flames. Approved Document B (2019) requires that every storey with a floor level more than 11 m above ground level has an alternative means of escape. Some tall buildings have staircases installed on the outside to prevent people from getting stuck in the corridors and provide access to fresh air while they escape.

As it can be seen from Figure 2.6, there was only one set of stairs for evacuation in Grenfell Tower. It is common for these kinds of old tower

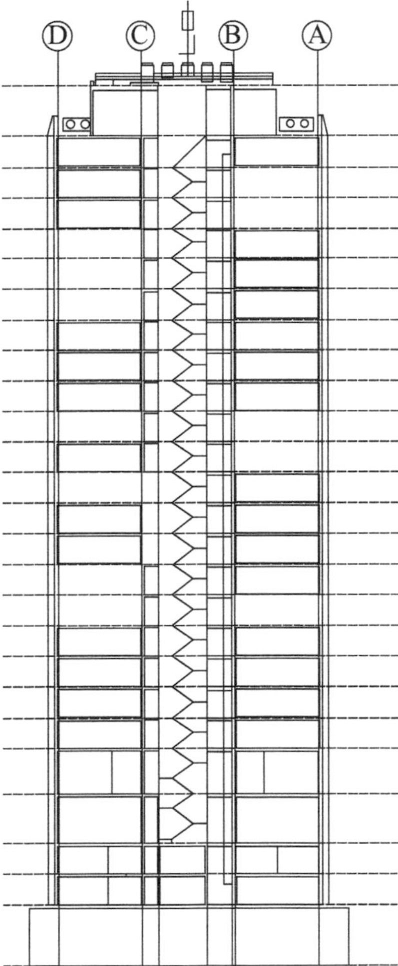

Figure 2.6 Evacuation route of Grenfell Tower.

blocks to have only one staircase, and no extra backup staircases available. Therefore, it slowed down the speed of the evacuation. It's clear that residents were not happy with the fire safety of the escape route from a blog posted before the disaster.

2.2.1.3 Collapse potential for Grenfell Tower

According to reports, Grenfell Tower is a concrete-framed block, which has high rate of fire resistance. While steel structural members can buckle in high temperatures, concrete structures can help to prevent the collapse of a building in case of fire, as well as making it safer to use helicopters—which can dump up to 9,842 litres of water at a time—to extinguish the blaze.

2.2.1.4 Major findings from Interim Report of British Research Establishment (2017)

The following major findings are reported from the report of British Research Establishment (2017):

1. The aluminum panels and insulation were used in the façade.
2. There was a compartmentation problem, i.e., lack of fire doors.
3. The new windows were too narrow.
4. The cavity barriers were of insufficient size.

Findings 1 and 2 have been discussed in the previous section. For finding 3, the UK regulation Approved Document B: Volume 1 (2019) makes different recommendations corresponding to the height of the building: a window might be an appropriate means of escape from a flat located on an upper level of a building (no higher than 4.5 m from ground level), but flats at a higher elevation would require alternative forms of emergency exits. Officially, as an old building, Grenfell Tower did not follow this regulation.

For finding 4, a cavity barrier can be a roof void barrier, underfloor cavity barrier suited to IT suites and offices with raised access flooring and an edge of slab fire protection detail between the building facade and the floor slab typically found in high-rise residential buildings. Cavity barrier is a vital fire seal of buildings, which in many cases is unseen but plays a vital role in containing fire and smoke within cavities at 20 m divisions.

Approved Document B (2019) is a fire safety regulation issued by the UK government, which will be introduced in the next section. From the investigation, questions have been raised about the state of fire safety regulation in England and Wales, particularly for high-rise buildings. Following allegations that the cladding used on Grenfell Tower may have contributed to the rapid spread of the fire, more stringent regulations were introduced in the new version of Approved Document B (2019), which will be introduced in Chapter 5.

2.2.2 Twin Tower

It is well known that on September 11, 2001, the Twin Towers (referred to as WTC1 and WTC2, respectively) were collapsed due to aircrafts crashing. However, the aftermath investigation by National Institute of Standards and Technology (NIST, 2005) shows that the collapse of the buildings is not due to the impact load from the aircrafts but due to the fires.

As shown in Figure 2.7 and Figure 2.8, the towers of WTC1 and WTC2 were designed with closely spaced steel mega columns at perimeter and steel cores in the center to provide robust stability system for such tall buildings. In between the perimeter and the center, there is a large column-free space which was bridged by prefabricated floor trusses. The whole structural system of the buildings can be simplified as shown in Figure 2.9. Apart from this so-called "Tube in Tube" system, the towers also used the conventional outrigger truss (also called Hat Truss) between the 107th and 110th floors

Figure 2.7 Structural system of WTC1 and WTC2. (Usrlman, the copyright holder of this work, release this work into the public domain, https://commons.wikimedia.org/wiki/File:WTC_bathtub_east.JPG.)

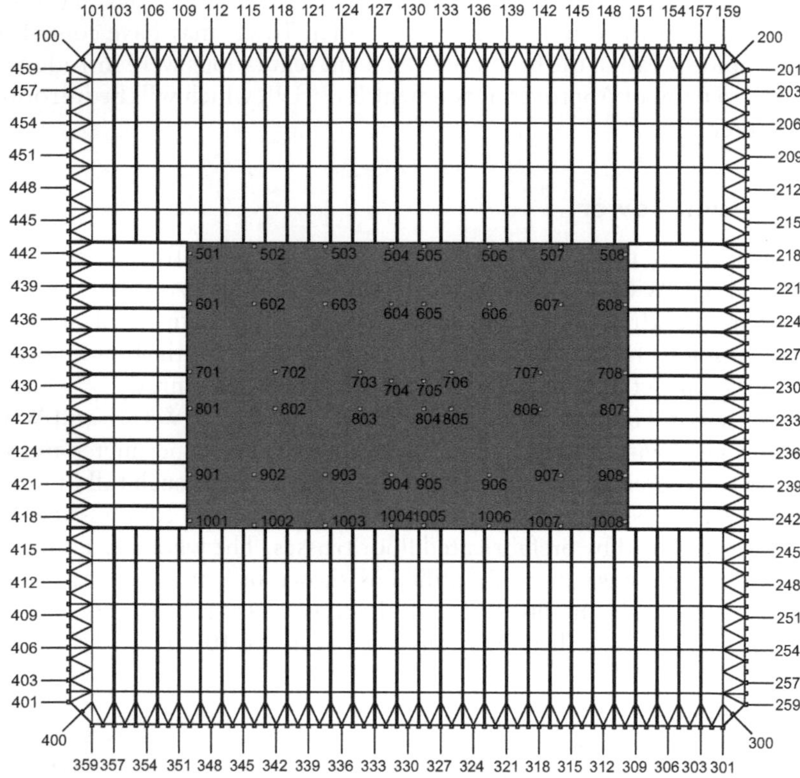

Figure 2.8 Typical floor layout for World Trade Center 1. (Reprinted from National Institute of Standards and Technology (2005), Federal building and fire safety investigation of the World Trade Center Disaster: Final report on the collapse of the World Trade Center Towers (U.S. Department of Commerce, Washington, DC), NIST NCSTAR 1, Figure 1–3, p. 7. https://doi.org/10.6028/NIST.NCSTAR.1.)

to further strengthen the cores. These special structural configurations guarantee a very strong structural system to resist terrific lateral load such as wind or earthquake. Therefore, it is fully capable to resist the impact load caused by the crashing of the aircrafts. For fire safety reason, the whole steel frame system including steel core and perimeter columns was also protected with sprayed-on fire-resistant material which can protect the structural members from fire. However, disengagement of the fire protections was reported by NIST (2005).

The report from NIST (2005) shows that after the aircrafts with full tanks of fuel hit WTC1 and WTC2, the explosion caused fire inside the two buildings. When the fire protection ceased to protect the structural members from the fire, the fire caused the sagging of the floor trusses, which further caused inward pulling of the perimeter columns. "This led to the inward bowing of

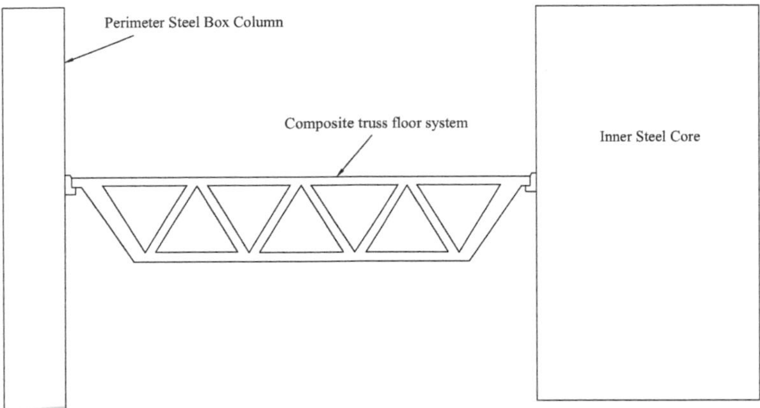

Figure 2.9 Schematic drawing of the structural system of Twin Towers.

the perimeter columns and failure of the south face of WTC1 and the east face of WTC2, initiating the collapse of each of the towers" (NIST, 2005). A 2D model was built by Usmani et al. (2003) to simulate the collapse mechanism of the Twin Tower, as shown in Figure 2.10.

It can be seen that the bowing of the floor caused buckling and failure of the perimeter columns as indicated in NIST (2015). It also clearly shows the progression of the failure at different times. As time passed by, the bowing of the floor and subsequent failure of the columns caused the overall failure of several storeys. When these storeys gave up, the storeys above them started to free-fall. It further increased the load for the storeys at the lower level, thus causing a so-called progressive collapse as shown in Figure 2.11.

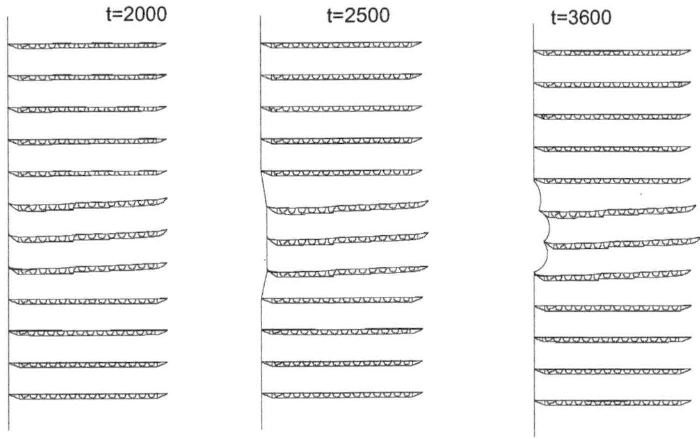

Figure 2.10 Collapse mechanism of Twin Towers (Usmani et al., 2003).

Figure 2.11 The process of progressive collapse in WTC1.

2.2.3 World Trade Center 7

World Trade Center building 7 is another tall building that collapsed due to fire. It is a 47-storey commercial building located next to the Twin Towers. As shown in Figure 2.12, the WTC7 is in irregular trapezoid shape. It is a steel composite-framed building. The main lateral stability is through steel moment connection frames with a center rectangular building core comprising 21 steel internal columns.

According to the investigation report of NIST (2008), the collapse of WTC7 was mainly due to the fire ignited as a result of the debris from the collapse of WTC1. There were both passive and active fire protection systems in WTC7 (NIST, 2008); the passive fire protection system is the sprayed fire-resistive material (SFRM) on the structural steel and metal decking for the floors. The active fire protection system consists of sprinklers inside the buildings; however, it was not functioning as the main water supply was cut off during the accident (NIST, 2008). The fire started to propagate inside the building, causing the buckling of columns, resulting in the failure of the floor above and further failure of the adjacent columns in the horizontal direction which triggered the progressive collapse of the whole buildings.

2.2.4 Other fire incidents of tall buildings

2.2.4.1 First Interstate Bank building in Los Angles

This is a 62-storey building using steel composite structural system. Fire started on level 12 and spread to four floors above. The building's steel beams and columns were fire protected; however, no fire compartment was designed. Therefore, there were no compartment walls and compartment floors. let alone fire stops. This caused nearly five storeys to burn out and partial building collapse.

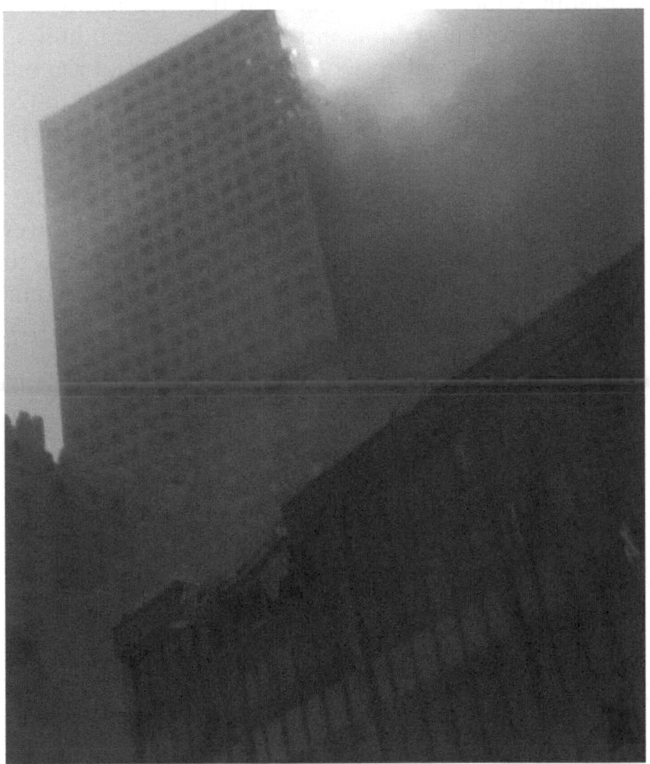

Figure 2.12 WTC7 in fire.(http://upload.wikimedia.org/wikipedia/commons/thumb/0/0e/Wtc7onfire.jpg/511px-Wtc7onfire.jpg, Courtesy of the Prints and Photographs Division. Library of Congress, public domain.)

2.2.4.2 *Plasco shopping center, Iran*

It is a 17-storey residential and commercial building using steel composite structural system, but no sprinkler or any other fire protection measures were adopted. Fire started on level 10 and spread to upper floors, causing the entire building to collapse (BBC, 2017).

2.2.4.3 *Faculty of Architecture Building, Delft University*

The fire started in a coffee-vending machine on the sixth floor of the 13-storey Faculty of Architecture Building at the Delft University of Technology, Delft, the Netherlands. The building had a reinforced concrete and steel structure consisting of a combination of six, three-storey structures, which effectively served as a podium, with a 13-storey tower located above. Although all occupants of the building were evacuated safely, the fire spread rapidly which severely impacted the operations of fire department, and thus the fire burned uncontrollably for several hours, eventually resulting in the structural collapse of a major portion of the building. The investigation of Meacham et al. (2010) and Engelhardt et al. (2013) shows the possibility of fire-induced spalling in RC columns leading to structural collapse due to excessive deflection of slabs.

This is a very rare example of concrete-building collapse due to fire. As it is widely known, the concrete is good in fire resistance. However, the fire burned for several hours before its extinction, which made the concrete to spall. The spalling can further induce the failure of the reinforcement, which will be explained in detail in Chapter 4.

2.2.4.4 *Windsor Tower, Spain*

It is a 32-storey building having a steel frame with concrete core wall as the major lateral stability system. Fire started on level 12 during the construction stage, causing partial collapse of the building. As the building was in the construction stage when the fire started, the sprinkler system and fire protection for exterior columns were not completely finished (Scoss Failure Data Sheet, 2008).

2.2.5 Cardington fire test

The Cardington fire test (British Research Establishment, 1999) is the first full-scale fire test for a multistorey building in the history. Fire tests were performed on an eight-storey typical braced steel office building at Cardington in the UK. As shown in Figure 2.13, a real Range Rover car was parked beside the building as an indication of the real size of the building tested in fire. The Cardington fire test is a milestone in fire safety design, as it clearly discovered the failure modes of structural components in case of fire, and it laid a foundation for modern structural fire design.

2.2.5.1 *Introduction of the test*

Figure 2.14 is a plan layout indicating the locations of different fire tests performed. Altogether six tests representing different scenarios were performed.

2.2.5.1.1 Test 1—single secondary beam— gas-fired furnace

As shown in Figure 2.14, Test 1 was a single secondary beam test. The beam and supported slab were heated up to 800°C–900°C using a gas furnace with the connections remaining at ambient temperature.

Figure 2.13 Full-scale multistorey building used for Cardington tests (British Research Establishment, 1999).

2.2.5.1.2 Test 2—plane frame test—gas-fired furnace

As shown in Figure 2.14, Test 2 was designed to investigate primary beams and columns along gridline B which supported the fourth floor. The primary and secondary beams and top columns were left unprotected. They were heated using gas furnace.

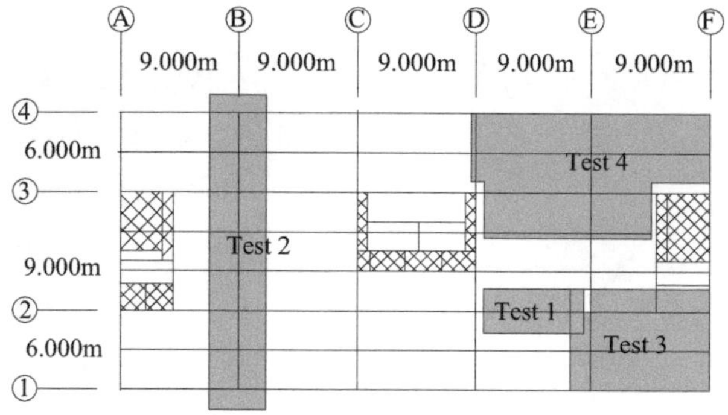

Figure 2.14 Floor layout and location of Cardington fire tests (British Research Establishment, 1999).

2.2.5.1.3 Tests 3 and 4—corner test—timber cribs = 45 kg/m²

As shown in Figure 2.14, both Tests 3 and 4 are corner compartment tests to investigate the composite floor system and membrane effect. They were heated up using wood cribs. It is equivalent to 45 kg/m^2 wood in fire.

In Test 3, all structural members were left unprotected apart from columns, column-to-beam connections, and external perimeter beams. The maximum recorded steel temperature was 935°C.

In Test 4, only columns were protected. Windows and doors were closed to simulate the fuel-controlled fire development, as the level of oxygen was restricted. It was recorded that the maximum steel temperature was 903°C.

2.2.5.1.4 Test 5—large compartment test

Test 5 was a large compartment test (340 m^2) which was conducted between the second and third floors. A fire resistance wall was constructed along the full width of the building. All the steel beams were left unprotected. The maximum atmosphere temperature and steel temperature recorded were 746°C and 691°C, respectively. Many beam-to-beam connections were found to have locally buckled, and many end-plate connections fractured down one side after cooling.

2.2.5.1.5 Test 6—demonstration test—office furniture ~45 kg/m² wood

Test 6 is to simulate the fire that occurs in an office; to closely simulate office environment, old office furniture was used as fuel in an open compartment of 135 m^2. It is equivalent to 45 kg/m^2 wood in fire. Only columns and beam-to-column connections were protected. The maximum steel temperature was 1,150°C.

2.2.5.2 *Failure modes for buildings in fire*

The Cardington fire test helps us to fully understand the failure modes of a multistorey building in fire, which provides guidance for structural fire design of tall buildings. Four major types of failure modes were observed during the full-scale fire tests. They are introduced in the following.

2.2.5.2.1 *Beam buckling and yielding*

As can be seen from Figure 2.15, during Test 1, local buckling at lower flange at the ends of the beam due to the restraining forces occurred due to thermal expansion against the web of the column section. Yielding at both ends of the test beam was also observed during the experiment. In Test 4, distortional buckling was observed along most of the beam length.

2.2.5.2.2 *Column buckling and yielding*

As shown in Figure 2.16, during Test 2, it is observed that the exposed parts of the columns squashed at approximately 670°C. This may lead to local collapse. Therefore, it is suggested by British Research Establishment (1999, 2004) that the columns should be fully fire protected along the entire length.

In Test 3, extensive buckling was noticed at beam-to-column connections. The end of an internal secondary beam which was connected to a primary beam buckled locally due to axial restraint from adjacent members. However, no local buckling occurred at the other end of the beam

Figure 2.15 Beam buckling (British Research Establishment, 1999).

Figure 2.16 Column buckling and yielding (British Research Establishment, 1999).

which was connected to an external beam. This was because the thermal expansion of the secondary beam caused the external beam to twist, resulting in insufficient restraint to cause local buckling.

2.2.5.2.3 Connection failure

In Test 3, it is also noticed that the bolts at the fin-plates connection were sheared off as shown in Figure 2.17.

Figure 2.17 Connection failure (British Research Establishment, 1999).

Figure 2.18 Slab failure (British Research Establishment, 1999).

2.2.5.2.4 Slab failure

During Test 6, it is found that no signs of failure were observed, but there were extensive cracking formed near the column zone during the latter phase of cooling as shown in Figure 2.18.

2.2.6 Discussion

From the introduction of the above incidents and Cardington fire tests, it can be seen that for the two steel framed buildings, WTC1 and WTC7 both collapsed. However, the concrete building Grenfell Tower did not. Though the concrete building in Delft University of Technology collapsed due to fire, it is quite rare. It can be seen that due to its fire resistance feature, concrete is the best option in the fire safety design. However, a pure concrete structure is not feasible. Most of the buildings are still steel framed. Therefore, fully understanding the design strategy and design method of tall buildings in fire is essential for designers.

In a tall building design, the staircase situated in the core of the buildings is one of the major evacuation routes for the occupants, which should be kept free of flame and smoke in the event of fire. Therefore, after the 9/11 incident, almost all the newly built tall buildings use concrete as the major material of core, and steel cores are not used for tall buildings.

2.3 CURRENT DESIGN GUIDANCE AND REGULATIONS TO FIRE SAFETY IN HIGH-RISE BUILDINGS

In the current design practice, there are several design codes and regulations for fire safety design used worldwide. A brief introduction of this guidance will be made in this section.

2.3.1 British design guidance and regulations

2.3.1.1 Building Regulations 2010—Approved Document B

The British government issued a series of approved documents that give practical guidance about how to meet the requirements of the Building Regulations 2010 for England. Approved Document B (2019) is a particular building regulation document that provides guidance for fire safeties of common buildings. It set out minimum standards in respect of health and safety which must be met when constructing new buildings. The fire design regulations in the UK take a dual approach: imposing requirements on new buildings (or those which undergo material alterations) and a separate set of requirements for existing buildings. In the case of existing multiple-occupancy buildings, obligations to ensure fire safety are imposed on both the "responsible persons" for a building, which can encompass owners, landlords, and managing agents, and also social housing landlords such as local authorities, housing associations, and any managing agents or tenants' association with delegated responsibilities. It comprises five major parts as follows.

2.3.1.1.1 Requirement B1: means of warning and escape

This part stipulates the regulations on fire detection and alarm system and the means of escapes from different levels of the buildings.

> *Requirement "Means of warning and escape*
>
> B1. The building shall be designed and constructed so that there are appropriate provisions for the early warning of fire, and appropriate

means of escape in case of fire from the building to a place of safety outside the building capable of being safely and effectively used at all material times."

2.3.1.1.2 Requirement B2: internal fire spread—linings

This part stipulates the regulations on a restricted spread of fire over internal linings. The building fabric should make a limited contribution to fire growth, including a low rate of heat release.

> *Requirement: "Internal fire spread (linings)*
>
> B2. (1) To inhibit the spread of fire within the building, the internal linings shall—
>
> (a) adequately resist the spread of flame over their surfaces; and
>
> (b) have, if ignited, either a rate of heat release or a rate of fire growth, which is reasonable in the circumstances.
>
> (2) In this paragraph 'internal linings' means the materials or products used in lining any partition, wall, ceiling or other internal structure."

2.3.1.1.3 Requirement B3: internal fire spread—structure

This part stipulates the regulations on the minimum fire resistance for the load-bearing structural elements such as frames, beams, columns, and load-bearing walls. It also stipulates on the design of the compartment.

> *Requirement "Internal fire spread (structure)*
>
> B3. (1) The building shall be designed and constructed so that, in the event of fire, its stability will be maintained for a reasonable period
>
> (2) A wall common to two or more buildings shall be designed and constructed so that it adequately resists the spread of fire between those buildings. For the purposes of this sub-paragraph a house in a terrace and a semi-detached house are each to be treated as a separate building.
>
> (3) Where reasonably necessary to inhibit the spread of fire within the building, measures shall be taken, to an extent appropriate to the size and intended use of the building, comprising either or both of the following—
>
> (a) sub-division of the building with fire-resisting construction.
>
> (b) installation of suitable automatic fire suppression systems.
>
> (4) The building shall be designed and constructed so that the unseen spread of fire and smoke within concealed spaces in its structure and fabric is inhibited."

2.3.1.1.4 Requirement B4: external fire spread

This part stipulates the regulations on resisting fire spread over external walls.

> *Requirement "External fire spread*
>
> B4. (1) The external walls of the building shall adequately resist the spread of fire over the walls and from one building to another, having regard to the height, use and position of the building.
>
> (2) The roof of the building shall adequately resist the spread of fire over the roof and from one building to another, having regard to the use and position of the building."

2.3.1.1.5 Requirement B5: access and facilities for the fire service

This part stipulates the regulations on access and facilities for the fire service.

> *Requirement "Access and facilities for the fire service*
>
> B5. (1) The building shall be designed and constructed so as to provide reasonable facilities to assist fire fighters in the protection of life.
>
> (2) Reasonable provision shall be made within the site of the building to enable fire appliances to gain access to the building."

2.3.1.1.6 Summary

It is important to note that the UK Building Regulations do not specify how these standards should be met, but how compliance achieved. There is no legal requirement to implement the guidance, provided that the minimum standards are met. Approved Document B, which has separate volumes for dwelling houses (Volume 1) and for other buildings (Volume 2).

Part B also requires that buildings be constructed in a manner to limit both internal and external fire spreads. Approved Document B provides prescriptive examples of how this can be achieved, but as noted above, compliance with Approved Document B is not mandatory. The penalty for contravention of the Building Regulations is an unlimited fine.

2.3.1.2 The FSO and Housing Act 2004

For existing buildings, there are other two regulations in UK which can be referred to: one is Fire Safety Order 2005 (FSO), which imposes duties on individuals in control of the building, the other one is Housing Act 2004 (the Act), which imposes monitoring duties on local authorities to take enforcement action against those in control of the building.

2.3.1.2.1 FSO

FSO imposes a duty on the "responsible person" to implement fire safety measures including undertaking a risk assessment; making fire safety arrangements; ensuring that a premise is equipped with fire-fighting equipment, fire-detection equipment, and emergency routes; and establishing procedures such as fire safety drills a "responsible person" will be a freehold owner or landlord but may also include a "residential management company" or a "right to manage company." Fire risk assessments should be conducted on a periodic basis to evaluate the risk and determine the appropriate fire safety measures to be implemented and maintained.

For tall buildings, the FSO only applies to the common parts. Responsible persons must ensure that tenant activities do not compromise the safety of the common parts (e.g. by placing obstructions/flammable objects in corridors or blocking fire-escape routes). Landlords should take appropriate action to minimize such risks, for example, by placing signs in prominent places instructing tenants to keep the common parts free from any obstruction and/or flammable objects. Landlords need to be mindful of these risks when drafting tenancy repair and maintenance obligations and should also ensure that regular checks are carried out/safety audits conducted on a routine basis. Certain repair works, including the replacement of fire doors, may constitute "material alterations" and must also comply with the Building Regulations.

Failure to provide adequate fire safety measures is a criminal offence. If the failure places one or more people at risk of death or serious injury, it can be punishable by an unlimited fine and/or up to 2 years' imprisonment. The Fire and Rescue Authorities have the responsibility for the enforcement of the FSO and will work in conjunction with responsible persons to monitor the safety of common parts of relevant buildings.

2.3.1.2.2 Housing Act 2004 (the Act)

The Act imposes a duty on local authorities to keep the housing conditions in their area under review. Local authorities may inspect the common parts of residential buildings, where they consider it appropriate to do so. In reviewing the state of the building, they must consider the 29 hazards prescribed by the Act, which include fire, noise, and structural collapse.

The local authority is required to take enforcement action where a Category 1 hazard (most serious) is identified and may do so at its discretion in respect of a Category 2 hazard. Such enforcement action can include improvement notices, emergency prohibition orders (preventing specified uses of the property), and demolition orders.

Under this Act, the Local Government Group publishes fire safety guidance to assist local authorities in understanding their responsibilities.

2.3.1.3 BS 7974:2019

BSI (2019) BS7974: 2019, "Code of Practice on Application of Fire Safety Engineering Principles to the Design of Buildings," is a British Standard providing a framework for designs that protect people, property, and the environment from fire. It contains guidance and information on how to undertake quantitative and detailed analyses of specific aspects of the design. They don't preclude the use of appropriate methods and data from other sources. It includes below parts:

Part 1: Initiation and development of fire within the enclosure of origin
Part 2: Spread of smoke and toxic gases within and beyond the enclosure of origin
Part 3: Structural response and fire spread beyond the enclosure of origin
Part 4: Detection of fire and activation of fire protection systems
Part 5: Fire service intervention
Part 6: Evacuation
Part 7: Probabilistic fire risk assessment

The primary users will be fire safety engineering practitioners. Other users include members of the fire and rescue service, structural engineers, architects, government departments, universities (for teaching and research), regulators, and related industries such as insurers and systems engineers.

2.3.1.3.1 PD 7974-3:2019

BSI PD7974-3 (2019), "Application of fire safety engineering principles to the design of buildings—Part 3: Structural response and fire spread beyond the enclosure of origin (Sub-system 3)," is primarily used for Structural Engineers in structural fire design.

PD 7974-3 (2019) considers the following issues:

- The conditions within a fire enclosure and their potential to cause the fire to spread by way of recognized mechanisms and routes.
- The thermal and mechanical responses of the enclosure boundaries and its structure to the fire conditions.
- The impact of these anticipated thermal and mechanical responses on adjacent enclosures and spaces.
- The structural responses of load-bearing elements and their effect on structural stability, load transfer, and acceptable damage.

2.3.1.4 BS 9999:2017

BS 9999 (2017), "Code of practice for fire safety in the design, management and use of buildings," gives recommendations and guidance on the design, management, and use of buildings to achieve reasonable standards intended to safeguard the lives of building occupants and fire fighters. It also provides guidance on the ongoing management of fire safety within a building throughout its life cycle, including guidance for designers to ensure that the overall design of a building assists and enhances the management of fire safety. It is applicable to the design of new buildings and also to alterations, extensions, and changes of use of an existing building. It is not applicable to individual dwelling houses and might have only limited applicability to certain specialist buildings and areas of buildings (e.g. hospitals and areas of lawful detention).

The primary users will be architects, fire safety engineers, fire risk assessors, building control, installers of fire and smoke alarms, sprinklers, and smoke and heat control systems.

2.3.1.5 BS 5950-8:2003

BS5950-8, British Standards Institution (2003), "The structural use of steelwork in buildings, Part 8: code of practice for fire-resistant design," gives recommendations for evaluating the fire resistance of steel structures. Methods are given for determining the thermal response of the structure and evaluating the protection to achieve the specified performance.

2.3.1.6 BS 476-20:1987

BS 476-20 (1987), "Incorporating Amendment No. 1. Fire tests on building materials and structures—Part 20: Method for determination of the fire resistance of elements of construction (general principles)," gives detailed specifications to determine the fire resistance of protected or unprotected elements of a building.

2.3.1.7 Design guidelines from IStructE and Steel Construction Institute

In the UK, the Institution of Structural Engineers and Steel Construction Institute also issued design guidelines for fire-resistant design. They particularly deal with structural fire design issues.

- IStructE (2007), Guide to the advanced fire safety engineering of structures, Institution of Structural Engineers, August 2007.

- The Steel Construction Institute, *Fire Resistant Design of Steel Structures, A Handbook to BS 5950: Part 8.*

2.3.2 Eurocode

Eurocode has probably the most detailed design guidance on fire safety design, including means of escape; fire spread; reaction to fire; the fire resistance of the structure in terms of resistance periods; the smoke and heat exhaust ventilation system; active firefighting measures such as hand extinguishers, smoke detectors, and sprinklers; and access for the fire brigade. The major design guidelines are as follows:

- EN 1991-1-2 (2002), Eurocode 1. Part 1–2: General actions—actions on structures exposed to fire.
- EN1992-1-2 (2004), Eurocode 2. Design of concrete structures, Parts 1–2: general rules—structural fire design.
- EN 1993-1-2 (2005), Eurocode 3. Design of steel structures, Parts 1–2: general rules. Structural fire design.
- BS EN 1994-1-2 (2005) Eurocode 4. Design of composite steel and concrete structures, Parts 1–2: general rules.

2.3.3 Guidelines from International Organization for Standardization

2.3.3.1 ISO 24679-1:2019(en)

ISO 24679-1:2019(en), "Fire safety engineering—Performance of structures in fire—Part 1: General," is a standard issued by International Organization for Standardization. It provides a methodology for assessing the performance of structures exposed to a real fire. It provides a performance-based methodology for engineers to assess the level of fire safety of new or existing structures. The fire safety of structures is evaluated through an engineering approach based on the quantification of the behavior of a structure for the purpose of meeting fire safety objectives and can cover the entire time history of a real fire (including the cooling phase) and its consequences related to fire safety objectives such as life safety and property protection.

2.3.3.2 ISO 16730-1 and ISO 16733-1

ISO 16730-1, "Fire safety engineering—Procedures and requirements for verification and validation of calculation methods—Part 1: General," and ISO 16733-1, "Fire safety engineering—Selection of design fire scenarios and design fires—Part 1: Selection of design fire scenarios," are the other two guidelines that deal with the calculation methods and selection of fire scenarios in fire safety engineering.

2.3.3.3 ISO 834-1:1999

International Organization for Standardization issued has another design code as well. ISO 834-1 (1999), "Fire-resistance tests—Elements of building construction—Part 1: General requirements," is a similar code to BS 476-20, which deals with the guideline on fire testing.

2.3.4 US design guidance

In the US, fire protection is impacted by a number of codes and standards. The most frequently used codes and standards are issued by the following professional bodies:

- National Fire Protection Association (NFPA)
- International Code Council
- American Society for Testing and Materials (ASTM) publishes several fire protection-related standards through its E-5 committee
- American Society of Civil Engineers (ASCE).

There are several codes and standards published by these professional bodies. They will be briefly introduced here.

2.3.4.1 National Fire Protection Association

NFPA has published several standards for fire safety design such as:

- NFPA 13, Standard for the Installation of Sprinkler Systems
- NFPA 72, National Fire Alarm Code
- NFPA 1, Uniform Fire Code.

NFPA has also published several codes and standards that cover specific aspects of fire protection and fire-related hazards.

2.3.4.2 International Code Council—International Fire Code® (IFC®)

The International Fire Code (2018) establishes minimum regulations for fire prevention and fire protection systems using prescriptive and performance-related provisions. This code addresses extraordinary fire risks in existing buildings with retrospective requirements, but only in this limited area is there a need for alterations, as long as the building and its occupancies comply with reasonable fire-prevention provisions.

2.3.4.3 American Society for Testing and Materials (ASTM)

ASTM has issued several standards related to fire safety as follows:

2.3.4.3.1 Specification for fire resistance

ASTM E119-19, "Standard Test Methods for Fire Tests of Building Construction and Materials," specifies test methods to evaluate the duration for which the types of building elements in fire.

ASTM E2748-12a (2017), "Standard Guide for Fire-Resistance Experiments," specifies methods and procedures set forth in this guide related to the conduct and reporting of fire-resistance tests obtained from particular fire-resistance specimens tested using conditions different from those addressed by Test Methods E119.

2.3.4.3.2 Specification for fire safety engineering

- E1355-12 (2018), Standard Guide for Evaluating the Predictive Capability of Deterministic Fire Models
- E1546-15 Standard Guide for Development of Fire-Hazard-Assessment Standards
- E1591-13 Standard Guide for Obtaining Data for Fire Growth Models
- E1776-16 Standard Guide for Development of Fire-Risk-Assessment Standards
- E3020-16a Standard Practice for Ignition Sources.

2.3.4.4 American Society of Civil Engineers

ASCE (1992), ASCE Manuals and Reports on Engineering Practice No. 78, Structural Fire Protection, Prepared by the ASCE Committee on Fire Protection Structural Division American Society of Civil Engineers, is a manual intended to provide a basis for calculation of the fire resistance of structural members. It not only focuses on design guidelines for structural fire safety design, but it also provides information on current techniques and developments to improve fire safety in buildings. It covers fire severity, response of various materials such as concrete and steel in fire, and fire protection.

The manual consists of two parts: the objective of Part 1, consisting of Chapters 1–3, is to introduce the subject matter to the building design practitioner who has had no experience with fire other than in work with building codes. The material in this part is mainly descriptive.

In Part 2, which consists of Chapters 4 and 5, the technical basis of the materials in Part 1 is described. This will enable those interested to obtain more knowledge about the background of the materials in Part 1.

2.3.4.5 *Federal Standards and Guidelines*

There are also some Federal Standards and Guidelines.

2.3.4.5.1 *Department of Defense (DOD)*

DOD UFC 3 600 01: Fire Protection Engineering for Facilities.

2.3.4.5.2 *General Services Administration (GSA)*

- PBS-P100 Facilities Standards for the Public Buildings Service
- "Fire Safety Retrofitting in Historic Buildings" by Advisory Council on Historic Preservation and General Services Administration, 1989.

2.3.5 Chinese design guidance

GB50016 (2014), Code for Fire Protection Design of Buildings published by National Standard of the People's Republic of China, is a design code for fire safety design of tall or multistorey buildings, tunnels, shopping centers, etc. It covers newly built buildings and extension or alteration to existing buildings. It stipulates the design requirements for compartmentation, evacuation, and specification of the dimension of individual member in a building to be able to comply with fire safety requirement.

CECS 392 (2014), Code specification for anti-collapse design of building structures by China Association for Engineering Construction Standardization, is a design specification for the collapse prevention of buildings. It comprises a part that specifies the guidance for fire-induced collapse, and the specification requires the structure to resist fire for a sufficiently long time without collapse. Three methods are introduced: the simplified component method, the alternative load path method, and the advanced analysis for entire fire process.

2.3.6 New Zealand code NZS 3404 Part 1:1997

The New Zealand steel code (NZS, 1997) includes a section for fire safety of essential steel elements of a structure. It specifies similar formulas of Eurocodes for the maximum temperature that an element can reach before it will no longer be able to carry the design load in fire condition and therefore fail, and the time until this temperature is reached. Other design tools include formulas to estimate the variation in mechanical properties of steel with temperature.

2.3.7 Australian code AS 4100:1998

The Australian Steel Code is similar to New Zealand Steel Code except for a few minor alterations such as the use of the exposed surface area to mass ratio (k_{sm}) instead of the section factor which is used in both New Zealand code NZS 3404. 1.5.3 and Eurocode ENV 1993-1-2.

2.4 BASIC PRINCIPLES FOR FIRE SAFETY OF TALL BUILDINGS

As tall buildings become more complex with dramatic changes in building envelope and materials, it is vital to consider fire safety implications of new buildings or other construction or refurbishment projects at the concept design stage. A successful fire safety design requires an understanding of a wide range of issues and components, and the interactions between them. At all stages of the project design, the following factors need to be considered:

- Regulations compliance
- Fire detection and suppression
- Fire modeling, and risk assessments
- Heat transfer to the structure
- Materials fire rating
- Fire protection measures—active and passive
- Smoke movement and smoke and heat exhaust ventilation systems
- People movement and means of escape.

2.4.1 Main design objective

It should be bore in mind at the beginning of this book that the main objective of fire safety design is to save lives, not to prevent collapse of buildings. The primary objective is to reduce the potential for death or injury to the occupants of a building and others who may become involved, such as the fire and rescue services. It is also crucial to protect contents of the buildings and ensure that, as much as possible, the building can continue to function after the fire or that it can be repaired.

Therefore, in the event of an outbreak of fire, the load-bearing elements of a building should continue to function until all occupants have escaped, or been assisted to escape, from the building. In addition, to achieve this primary objective, effective compartmentation and evocation route, as well as other factors such as means of warning, should be guaranteed in the design.

2.4.2 Main design tasks

Based on the key objective, there are many aspects that need to be addressed when designing fire safety, and particularly the fire safety design for tall buildings encompasses a wide range of techniques addressing:

- Means of warning and escape
- Compartmentation and ventilation
- Structural fire design
- Smoke control, spread of smoke
- Active measures for fire containment and control (sprinklers)
- Fire safety management
- Human behavior in the event of a fire.

However, among them, the key design tasks for fire safety are compartmentation, evacuation, and structural fire design. These three factors affect each other; for example, when designing the evacuation route, the time of evacuation is affected by the time of failure of structural members in fire. Therefore, an effective fire protection design is also essential to evacuations. These design tasks and how to achieve them will be discussed in detail in Chapter 3.

2.4.3 Structural fire design

As the structural fire design is the key for fire safety design, it is worth a brief introduction at the beginning of this book. Structural fire design determines the thermal behaviors of the structural members and finds effective means of fire protections to satisfy fire safety design objectives.

First, the atmosphere temperature needs to be determined, and then, heat transfer to the structural elements must be calculated. Different levels of analysis can be used (design formula, simplified model, or sophistic finite element analysis). When the temperature field of the structure is obtained, from the combination of the mechanical and thermal loads in case of fire, the thermal behaviors the structural elements can be assessed, which allows for further assessment against a range of performance criteria in terms of deformation and structural damage.

The choice of performance for design purposes will be dependent on the consequences of failure and the function of the building. For certain high-profile multistorey buildings, this may mean that no structural failure is allowed to take place during the whole duration of the fire.

2.4.3.1 Key design tasks in structural fire design

For a structural fire design, there are two key design tasks: stability and integrity.

2.4.3.1.1 Stability

The overall structural stability of the building as well as the primary structural elements (columns, beams, connections, load-bearing walls) should continue to function to ensure sufficient time for the evacuation of all the occupants, or until extinction of the fire.

2.4.3.1.2 Ensure compartment integrity

In the event of fire, compartments should continue to be maintained integrated to limit the passage of smoke and flame. In a structural fire design, this is achieved primarily through the control of deflections of key compartment components such as slab and beam-supporting compartment wall.

2.4.3.2 Design approach

There are two major design approaches that can be used for a structural fire design, namely, prescriptive-based design and performance-based design.

2.4.3.2.1 Prescriptive-based design

This method is to set up safety factors by constraining design output to pre-established bounds; in other words, it is to design the structure based on fire rating of materials which is in compliance with a code-specified value. If a designer follows these rules, they will fall within the bounds, and the design can be finished.

2.4.3.2.2 Performance-based design

A designer needs to first understand the level of the performance that is expected (Custer and Meacham, 1997) and then satisfy this level in the design. It includes evaluating the strength and stiffness of the structural members for a particular designed fire, and thus achieving the stability of the structure.

2.4.3.3 *Pros and cons of the two design methods*

The constraints of the prescriptive design have been noticed by most of the practicing engineers. One of the famous examples is Grenfell Tower, which simply satisfies the bounds of the fire rating if materials do not necessarily satisfy the objective of fire safety. Therefore, it can be noticed that the current design codes in most of the countries are moving toward the latter approach. Through the investigation on WTC7, NIST NCSTAR (2008) also recommends that

> the fire resistance of structures can be enhanced by requiring a performance objective that building fires result in burnout without partial or global (total) collapse. The prescriptive design methods for determining the fire resistance rating of structural assemblies do not explicitly specify a performance objective. The rating resulting from current test methods indicates that the assembly continued to support its superimposed load during the test exposure without collapse, however it is collapse

These two design methods will be described in detail in Chapter 5.

2.4.4 Robustness of the structure in fire

Although the key design objective when buildings in fire is saving lives, and not to prevent collapse, when a building is in fire, if the building collapses, for sure it will affect the safety of lives. Therefore, the collapse of the building during the fire should be prevented. In Chapter 7, this topic will be discussed in detail.

2.4.5 Fire modeling

In a fire safety design, fire modeling is an important tool for effective delivery of feasible design measures. It comprises two major areas:

- Modeling the atmosphere temperature induced by fire.
- Modeling the thermal response of building elements (primarily load-bearing structural members or sometimes non-load-bearing members).

2.4.5.1 *Modeling the atmosphere temperature induced by fire*

There are several ways to model the atmosphere temperature during the fire such as CFD or zone model, filed model, or simple model (by using some simplified formula to calculate the atmosphere temperature, Eurocode adopted this approach).

2.4.5.2 Modeling the thermal response of load-bearing building elements

1. Simplified models

   If the heated temperature of the structural member is below the critical temperature, there is no failure, but if it is higher than the critical temperature, there is failure. It is a "pass or failure" criterion. The objective is then reached if the time to reach the failure is greater than the required natural fire exposure. This so-called critical temperature method is adopted in both British code and Eurocode. This method will be explained in detail in Chapter 4.
2. Finite element model

   Finite element modeling software (Fu, 2016) is required to model the response of the structural members in fire. The results are assessed generally in terms of deformation during the whole fire duration. The performance criteria (to measure at which level the objectives are fulfilled) in terms of deformation can be used for the assessment.

Another more stringent and complicated method is the so-called multi-physics thermal–mechanical coupled modeling, which is the most accurate way to assess the behaviors of the structural member under fire. However, it is computationally expensive, and thus it is not necessary in most of the cases.

2.4.5.3 Summary

As it can be seen, fire modeling is an important tool in fire safety design, and it will be discussed in detail in Chapter 6.

REFERENCES

AS 4100 (1998), Building code of Australia, steel structure, standards Australian.

ASCE (1992), ASCE Manuals and Reports on Engineering Practice No. 78: Structural fire protection, ASCE Committee on Fire Protection Structural Division.

ASTM E119-19, Standard test methods for fire tests of building construction and materials.

BBC (2017),Tehran Fire: Twenty Firemen killed as High-rise Collapses, https://www.bbc.co.uk/news/world-middle-east-38675628.

British Research Establishment (1999), *The Behaviour of Multi-Storey Steel Framed Buildings in Fire*. British Steel Plc Swinden Technology Center. ISBN 0900206500

British Research Establishment (2004), 'Client report: Results and observations from full-scale fire test at BRE Cardington, 16 January 2003 Client report number 215–741', February 2004 (Accessible from: http://www.mace.manchester.ac.uk/project/research /structures/strucfire/DataBase/TestData/default1.htm)

British Research Establishment (2017), Interim BRE global client report for Grenfell Tower:

BS 476-20 (1987), Incorporating Amendment No. 1. Fire tests on building materials and structures, Part 20: Method for determination of the fire resistance of elements of construction (general principles).

BS5950-8, British Standards Institution (2003), The structural use of steelwork in buildings, Part 8: Code of practice for fire resistant design.

BSI BS 9999 (2017), Code of practice for fire safety in the design, management and use of buildings, The British Standards Institution.

BSI 'PD7974-3 (2019), Application of fire safety engineering principles to the design of buildings, Part 3: Structural response and fire spread beyond the enclosure of origin (Sub-system 3)', The British Standards Institute.

BSI'BS7974 (2019), Code of practice on application of fire safety engineering principles to the design of buildings', The British Standards Institution.

CECS 392 (2014), Code Specification for anti-collapse design of building structures by China Association for Engineering Construction Standardization.

Custer, R., Meacham, B. J. (1997), Introduction to performance-based fire safety, Society of Fire Protection Engineers.

EN 1991-1-2 (2002), Eurocode 1. Actions on structures, Part 1–2: General actions. Actions on structures exposed to fire. Commission of the European communities

EN 1992-1-2 (2004), Eurocode 2. Design of concrete structures, Part 1–2: General rules. Structural fire design. Commission of the European communities.

EN 1993-1-2 (2005), Eurocode 3. Design of steel structures, Part 1–2: General rules. Structural fire design. Commission of the European communities.

EN 1994-1-2 (2005), Eurocode 4. Design of composite steel and concrete structures, Part 1–2: General rules. Structural fire design. Commission of the European communities.

Engelhardt, M., Meacham, B., Kodur, V., Kirk, A. (2013), Observations from the fire and collapse of the faculty of architecture building, Delft University of Technology, DOI: 10.1061/9780784412848.101, Structures Congress 2013

Fu, F. (2016), 3D finite element analysis of the whole-building behaviour of tall building in fire. *Advances in Computational Design*, 1(4), pp. 329–344.

Fu, F. (2017), Grenfell Tower disaster: How did the fire spread so quickly? *BBC Australia.*

GB50016 (2014), Code for Fire Protection Design of Buildings, National Standard of the People's Republic of China.

HM Government (2019), The Building Regulations 2010-Approved Document B, Volume 1 fire safety, Dwellings', HM Government.

HM Government (2019), The Building Regulations 2010-Approved Document B, Volume 2 fire safety, Buildings other than Dwellings', HM Government.

International code council (2018), International fire code. ISBN 978-1-60983-739-6, INTERNATIONAL CODE COUNCIL, INC. Date of First Publication: August 31, 2017, U.S.A.: Publications, 4051 Flossmoor Road, Country Club Hills, IL 6047.

ISO 16730-1, Fire safety engineering—Procedures and requirements for verification and validation of calculation methods, Part 1: General.

ISO 16733-1, Fire safety engineering—Selection of design fire scenarios and design fires, Part 1: Selection of design fire scenarios.

ISO 834-1 (1999), Fire-resistance tests—Elements of building construction, Part 1: General requirements, Edited by T. T. Lie.

ISO 24679-1 (2019), Fire safety engineering—Performance of structures in fire, Part 1: General

IStructE (2007, August), *Guide to the Advanced Fire Safety Engineering of Structures*, Institution of Structural Engineers, London.

Meacham, Brian J., Park, H., Engelhardt, M., Kirk, A., Kodur, V., van Straalen, I., Maljaars, J. P., van Weeren, K., de Feijter, R., Both, K. F. (2010), Fire and collapse, Faculty of Architecture building, Delft University of Technology: Data collection and preliminary analyses.

NIST (2007), Best practices for reducing the potential for progressive collapse in buildings, National Institute of Standards and Technology, Technology Administration, U.S. Department of Commerce.

NIST NCSTAR (2005, December), Federal building and fire safety investigation of the World Trade Center disaster, final report of the National Construction Safety Team on the collapses of the World Trade Center Towers.

NIST NCSTAR 1A (2008), Final report on the collapse of world trade center building 7, National Institute of standards and Technology, US department of commerce.

NZS 3404 Parts 1 and 2:1997, Steel Structures Standard.

Scoss Failure Data Sheet (2008), The Fire at the Torre Windsor Office Building, Madrid 2005.

The Steel Construction Institute, *Fire Resistant Design of Steel Structures*, A Handbook to BS 5950: Part 8

Usmani, A. S., Chung, Y. C., Torero, J. L. (2003), How did the WTC towers collapse: A new theory. *Fire Safety Journal*, 38(6), pp. 501–533.

Chapter 3

Fundamentals of fire and fire safety design

3.1 INTRODUCTION

In this chapter, the fundamental knowledge of fire and fire safety design will be explained. The characteristics of fire and its development are introduced at the beginning. Then, the key scenarios that affect the performance of the building members in fire—such as ventilation-controlled or fuel-controlled fire, and long-cool, short-hot fire—will be explained. In addition, the fundamentals of heat transfer, a process of the heating up of structural members due to fire, will be introduced. The basic structural fire design principles will also be explained. In fire safety design, most of the codes specify the fire resistance for building elements. The relevant information will be provided in the latter part of this chapter followed by the introduction of fire protection methods.

3.2 FIRE DEVELOPMENT PROCESS

As shown in Figure 3.1, the development process of a real fire in a confined area consists of below major phases:

- **Ignition**: ignition and smoldering of fire at very low temperatures with varying durations.
- **Growing phase or pre-flashover (localized fire)**: the duration of this phase depends mainly on the characteristics of the compartment. Still, the fire remains localized.
- **Flashover**: this phase is generally very short. The temperature sharply increases at this stage.
- **Post-flashover fire (fully developed)**: this phase corresponds to a generalized fire for which the duration depends on the fire load and the ventilation.
- **Decay phase**: the fire begins to decrease after burning all combustible materials completely.

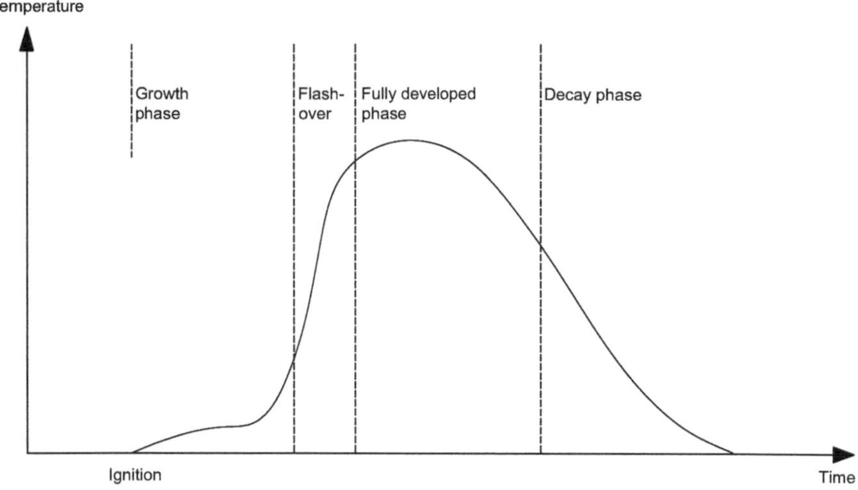

Figure 3.1 Temperature–time curve for fire development in a confined area.

3.3 DESIGN FIRE TEMPERATURE

In structural fire design, design fire temperature curves are used to represent the atmosphere temperature (compartment temperature). As shown in Figure 3.2, there are two main fire temperature curves from current design codes that the engineers can use directly in their structural fire analysis, namely, standard fire temperature and parametric fire temperature.

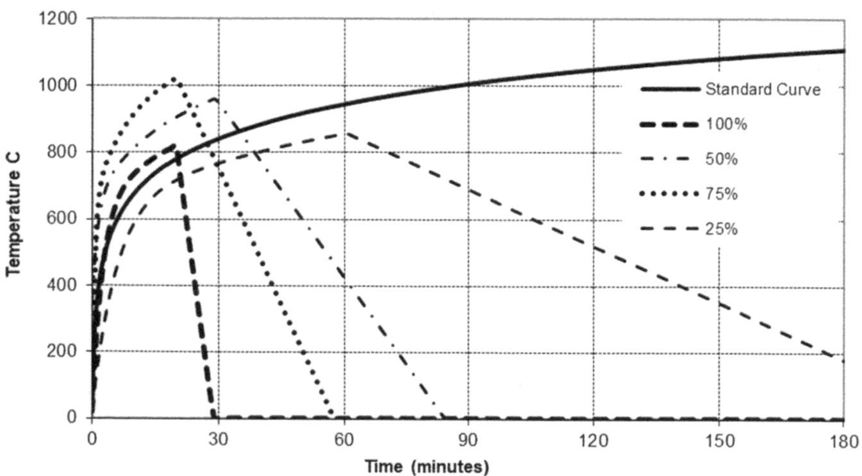

Figure 3.2 Standard fire time temperature and parametric fire temperature with different opening factors (570 MJ/m² fire load density).

3.1.1 Standard fire temperature–time curve

As shown in Figure 3.2, standard fire temperature curve is primarily used to define the heating condition for fire test on the structural members, but the cooling phase is not included. BS476: part20 (1987) gives the formula to calculate the temperature:

$$T = 345\log_{10}(8t+1)+20 \tag{3.1}$$

where
T is the mean furnace temperature (in °C),
t is the time (in min) up to a maximum of 360 min.

3.1.2 The parametric temperature–time curves

The parametric fire is defined in Annex A of EN1991-1-2: Eurocode 1; Part 1.2 (2002) (see Figure 3.2). The parametric temperature–time curves has both heating and cooling phase. It resembles the real atmosphere temperature Its heating phase given by EN1991-1-2: Eurocode 1; Part 1.2 (2002) are obtained as follows:

$$\Theta_g = 20+1325\left(1-0.324e^{-0.2t^*}-0.201e^{-1.7t^*}-0.472e^{-19t^*}\right) \tag{3.2}$$

where
Θ_g is the gas temperature in the fire compartment,

$$t^* = \Gamma t,$$

t = time,

$$\Gamma = [O/b]^2/[0.04/1160]^2,$$

O = opening factor,

$$O = \frac{A_v\sqrt{H_w}}{A_t}$$

where
A_t = total internal surface area of compartment [m^2]
A_v = area of ventilation [m^2]
H_w = height of openings [m]
b = thermal diffusivity, $100 \le b \le 2{,}000\left(\text{J/m}^2\text{s}^{1/2}\text{K}\right)$,

The maximum temperature Θ_{max} in the heating phase happens when $t^* = t^*_{max}$:

$$t^*_{max} = t_{max}\Gamma$$

$$t^*_{max} = t_{max} \bullet \Gamma$$

with

$$t_{\max} = \left(0.2\ 10^{-3}\frac{q_{t,d}}{O}\right)$$

$$t_{\max} = \left(0.2 \cdot 10^{-3} \bullet q_{t,d}/O\right)$$

or $t_{\lim}$.

$q_{t,d}$ is the design value of the fire load density related to the total surface area At of the enclosure:

$$q_{t,d} = q_{f,d} \times \frac{A_f}{A_t}\left[\text{MJ/m}^2\right]$$

The following limits should be observed:

$$50 \le q_{t,d} \le 1000\left[\text{MJ/m}^2\right]$$

$t_{\lim}$ in case of slow fire growth rate, $t_{\lim}$=25 min;
in case of medium fire growth rate, $t_{\lim}$=20 min;
in case of fast fire growth rate, $t_{\lim}$=15 min.

From Equation 3.2, it can be seen that the opening factor plays an important role in the atmosphere temperature. As it can be seen from Figure 3.2, under the same fire load density of 570 MJ/m^2, when the opening factor increases from 25% to 100%, the parametric fire temperature curves change significantly, the fire changes from "long-cool" to "short-hot." These two fire scenarios have distinctive effects on the response of structural members. This will be explained in detail Section 3.4.

3.1.3 Summary

From the above, it can be seen that, compared to standard fire curve, parametric fire curve has a cooling phase. The parametric fire curve more closely represents the real fire temperature than standard fire curve. For designing the fire resistance capacity of each individual member, the standard fire curve shall be used, as it gives more conservative results. For determining the behavior of a structural member, especially behavior of the whole-building or a frame, the parametric fire curve shall be used, as it gives more realistic fire temperature development in a compartment.

3.4 DESIGN FIRE IN A COMPARTMENT

Fire conditions depend on many factors such as the building's function (offices, car parks, etc.) and the materials (such as concrete or steel) used.

The complexity of tall buildings can cause different fire scenarios. For instance, depending on the compartmentation and structural layout of buildings, different fire temperatures can be reached. Different factors (such as fuel or ventilation) can also affect the development of the fire. The duration of the fire is determined by the opening factors and the type of fuels (such as the furniture and decorations).

As indicated in PD7974-part 3 (2019), the design fires in a compartment need to consider below key factors:

- Nature of the fire load.
- Fire load density.
- Thermal properties of the enclosure.
- Extent of ventilation.
- Ceiling height.

For determining fire conditions in each compartment, in PD7974-part 3(2019), three methods are given:

1. Standardized models with recognition of fire conditions and occupancy type.
2. Experimental data appropriate to the compartment and occupancy type.

   This method primarily uses the means of existing experimental tests data to determine the fire conditions. However, although it relies on a vast amount of test data, the required data may not be available for certain specific cases most of the time.
3. Engineering calculations based upon experimentally calibrated methods.

   This method primarily uses experimentally calibrated design rules to determine the fire conditions. It is simpler and more efficient for an engineer. It will be described in the following sections.

3.4.1 Characterization of compartment

For determining the design fire, the compartment needs to be characterized first. It comprises two major characterizations: enclosure and opening of the compartment.

3.4.1.1 Characterization of fire enclosure

In compartment design, compartment walls at the boundary of the compartment are purposely designed to prevent the spreading of fire. The compartment should be evaluated on fire conditions and the potential for fire spread. The fire resistance of compartment walls can be tested in a standard furnace.

In addition, in a building, all solid boundaries can slow down the spread of fire to some extent. The boundary members of a compartment are particularly important in terms of growth rate of the fire and the development of temperatures within the space. Therefore, fire conditions correlate with the capacity of initial enclosing surfaces to remain imperforate to the spread of fire.

For characterizing the enclosure of fire origin, PD7974 Part 3 (2019) offers the following guidance:

- When predicting pre-flashover fire conditions, the horizontal and vertical surfaces immediately surrounding the fire should be considered as the enclosure.
- The enclosure can also contain openings which, although immediately and directly open to the passage of fire and heat, may be characterized as part of the enclosing boundaries.
- After flashover has occurred, solid boundaries may be assumed to remain imperforate to fire as long as there are no openings created on their surfaces due to the mechanical force of the fire.
- As the definition of the enclosing surfaces is changed by the creation of openings, the fire conditions may need to be re-examined and fire spread routes re-evaluated.

3.4.1.2 Characterization of openings

Openings such as door, windows, and vents play an important role in fire development in a compartment, as they permit airflow to the fire and ventilation of heat. The characterization of openings includes their size, shape, and extent. Where a combination of fixed opening conditions is possible for the fire enclosure (e.g. some doors are open, and some doors are closed), options that are most conducive to fire spread should be considered. The assumption that all openings are initially open may not necessarily be the worst case. The following guidance is offered by PD7974 Part 3 (2019):

- Doors should be assumed open if the enclosure has no other openings.
- Doors should be assumed closed if the enclosure has other openings.
- All enclosure surfaces (including glazed openings) may be assumed to be imperforate for the duration of the fire.

3.4.1.3 Duration of fire to be adopted in design

The appropriate duration of the fire is influenced by the following parameters:

- Occupancy;
- Presence of automatic sprinkler system;

- Height of enclosure above ground level;
- Depth of enclosure below ground level.

3.4.2 Fuel-controlled and ventilation-controlled fire

In the fire development process, the amount of fuels (combustible material in a compartment) and the opening factor are the two key influential factors. As the amount of fuels dose not vary too much for certain types of buildings (either commercial or residential), the percentage of openings is the main variable for determining worst-case conditions of the compartment.

As it has been explained, flashover is the transition from the localized fire to the compartment room fire. However, for rooms with very large window openings, too much heat may be released through the windows, which leads to the occurrence of flashover. Even at the post-flashover stage, the rate of the combustion also depends on the size and shape of the ventilation openings.

A ventilation-controlled fire occurs when there is not enough air to support the complete combustion process of the fuel in a compartment. Therefore, the fire will extinct when the available oxygen runs out.

A fuel-controlled fire is where there is adequate amount of air but not enough fuel to support the combustion process of the fuel in the fire compartment. Therefore, the fire will extinct when all the fuels are burnt. As introduced in Chapter 2, Test 4 of Cardington tests is to simulate the fuel-controlled fire with all windows closed.

As can be seen in Equation 3.2, the opening factor O included in the equation is essential to determine whether a fire is fuel controlled or ventilation controlled. Large opening factor allows more oxygen to enter into the compartment, thus leading to fuel-controlled fire. Smaller opening factor allows less oxygen to enter into the compartment; therefore, ventilation-controlled fire is likely to happen.

3.4.3 Long-cool and short-hot fire

Depending on the opening factors, when exposed to fires, the structure responds in two distinct ways:

Under "short-hot" fire, the unprotected steel reaches the temperature similar to the fire temperature in atmosphere. However, for a concrete slab, due to its lower thermal conductivity, under short fire exposure, it reaches an average temperature only marginally higher than ambient conditions. So, steel and concrete material behave distinctively in "short-hot" fire. In addition, large deflections develop in a very short time. Research by Lamont et al. (2004) shows that large deflections develop in a very short time, which may result in early compartmentation failure.

Under "long-cool" fire, structural members experience much longer duration of heating, and so both steel and concrete members can reach considerably high average temperatures.

For steel composite structural members, in "short-hot" fire, the composite floor structure will experience higher temperature gradients across the cross-section. Therefore, thermal bowing is greater than that in the "long-cool" fire. The long duration of the "long-cool" fire results in higher temperatures in the concrete and the steel. Due to higher temperatures achieved in concrete slabs, there is much less tension in the slab with growing compression toward the end of heating.

Therefore, it can be concluded that in most of the cases, the worst-case fire scenario in terms of the structural response is often "short-hot" fire.

3.4.4 Fully developed fire

Fully developed fire is the stage after flashover where all combustibles are fully burnt.

For determining the atmosphere temperature for a fully developed fire, using parametric fire temperature from Eurocode is the simplest method. Zone model and CFD model can also provide more accurate results.

3.4.5 Localized fire

In large open-space buildings such as airport terminals, railway stations, large industrial halls, car park, and some commercial buildings, post-flashover fire is unlikely to occur. Here, fuel-controlled fire would be more likely. The common scenario in this type of plan layout is that the fire starts as a localized fire. As shown in Figure 3.3, a localized fire is a pre-flashover fire which is expected to burn locally, and the temperature of a structural member is not uniform along the plume.

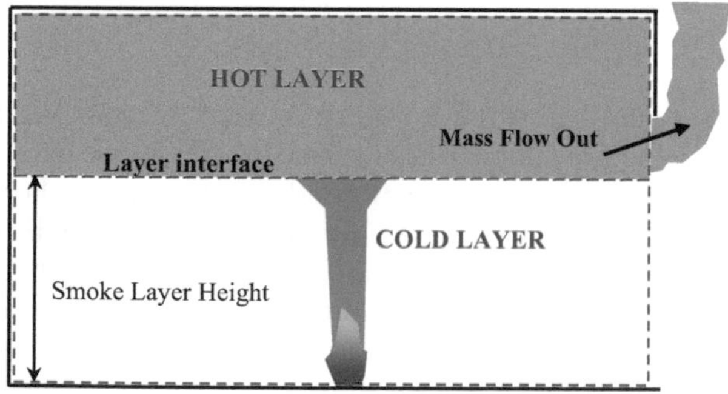

Figure 3.3 Schematic drawing for localized fires.

Eurocode assumes that the heating is uniform which in certain circumstance is not conservative, especially in the case of localized fires. High-intensity fire exposure effects may cause more severe damages to the structural members. The thermal response of such fires depends not only on the flame temperature but also on the magnitude and dimensions of the flames. Zhang et al. (2003) suggest that the failure mode of a beam may be different if it is exposed to a localized fire instead of the standard fire curve. Localized fire could still expose structures to severe thermal conditions, even though the mean temperature in the enclosure is low. It can significantly weaken the load-bearing capacity of a structural member in comparison to a uniform fire exposure.

For determining the atmosphere temperature for a localized fire, using a simplified plume model Eurocode 1 (EN 1991-1-2, 2002) is a common method, which will be introduced here. Other more accurate methods would be using one-zone model and CFD model, which will be introduced in Chapter 6.

3.4.5.1 Calculation of thermal action of a localized fire from Eurocode

EN 1991-1-2 (2002) specifies the thermal action of a localized fire with the consideration of the flame length.

The flame lengths L_f of a localized fire (see Figure C.1) is given by

$$L_f = -1.02D + 0.0148Q^{\frac{2}{5}} \tag{3.3}$$

- When the flame does not impact the ceiling of a compartment ($Lf<H$; see Figure C.1) or in case of fire in open air, the temperature $\Theta(z)$ in the plume along the symmetrical vertical flame axis is given by

$$\Theta(z) = 20 + 0.25(Q_c)^{-\frac{5}{3}} < 900 \tag{3.4}$$

where

D is the diameter of the fire [m], see Figure 3.4,

Q is the rate of heat release [W] of the fire,

Q_c is the convective part of the rate of heat release [W], with Q_c=0.8 Q by default,

z is the height [m] along the flame axis, see Figure 3.4,

H is the distance [m] between the fire source and the ceiling, see Figure 3.4.

- When the flame impacts the ceiling ($L_f > H$; see Figure 3.5), the heat flux $h\_$ [W/m^2] received by the fire-exposed unit surface area at the level of the ceiling is given by

$h\_=100{,}000$ if $y \leq 0.30$

$h\_=136{,}300-121{,}000\ y$ if $0.30<y<1{,}0$ (C.4)

$h\_=15{,}000\ y^{-3{,}7}$ if $y>1.0$

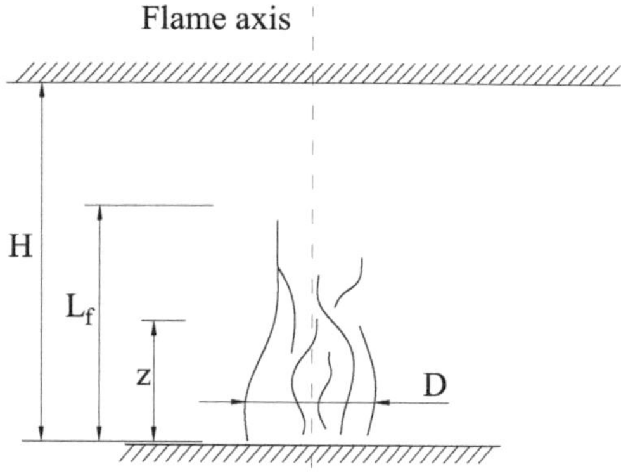

Figure 3.4 Localized fire (flame is not impacting the ceiling). (Figure C.1 of EN 1991-1-2 (2002), Eurocode 1. Actions on structures, Part 1-2: General actions. Actions on structures exposed to fire. Commission of the European communities 2002; permission to reproduce and extracts from British and ISO standards is granted by BSI. British Standards can be obtained in PDF or hard copy formats from the BSI online shop: www.bsigroup.com/Shop or by contacting BSI Customer Services for hardcopies only: Tel: +44 (0)20 8996 9001, Email: cservices@bsigroup.com.)

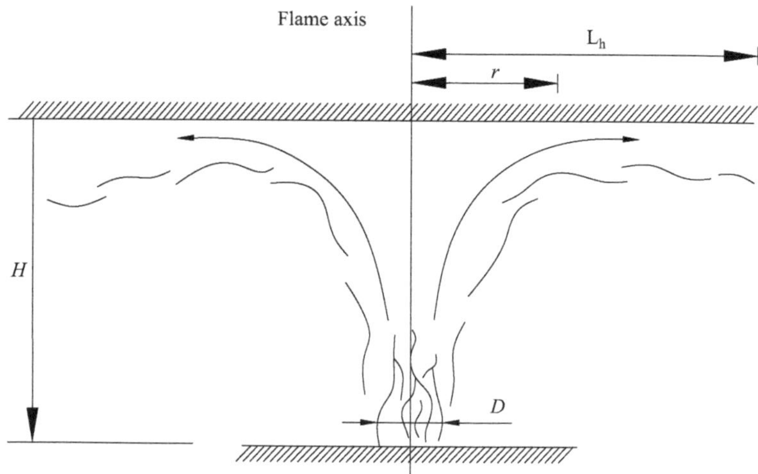

Figure 3.5 Localized fire (flame is impacting the ceiling). (Figure C.2 of EN 1991-1-2 (2002), Eurocode 1. Actions on structures, Part 1-2: General actions. Actions on structures exposed to fire. Commission of the European communities 2002; permission to reproduce and extracts from British and ISO standards is granted by BSI. British Standards can be obtained in PDF or hard copy formats from the BSI online shop: www.bsigroup.com/Shop or by contacting BSI Customer Services for hardcopies only: Tel: +44 (0)20 8996 9001, Email: cservices@bsigroup.com.)

3.4.6 Traveling fire

As introduced in the proceeding sections, in current fire safety design, homogeneous compartment temperature assumption is made. However, fire in large compartments such as tall buildings features in non-uniformity. The fire that burns locally may spread across entire floor plates over a period of time (Rein, 2007). This kind of fire scenario is called traveling fire. The experimental study of Horová (2013) shows that during traveling fire, a high degree of non-uniformity is observed in temperature. The maximum degree of heterogeneity is reached with a temperature difference up to 400°C. A temperature difference of up to 300°C can also be observed between the center and the corners of a fiber-reinforced concrete slab. As the fire spreads in the compartment, gas temperature under the ceiling fluctuates significantly, which affects structural member.

In real life, traveling fires have been observed in several fire incidents, for example, the World Trade Center Towers (NIST, 2005), and the Windsor Tower (Fletcher et al., 2006) in Madrid in 2005.

As explained by Ellobody and Bailey (2011), the peak temperature changes as the fire dynamic fields with cyclic heating and cooling can appear. The pattern of cyclic temperature changes can cause cyclic deflection, leading to a dangerous scenario.

3.4.7 Fire scenarios for tall buildings

As it has been discussed in the proceeding sections, localized and traveling fires dominate fire scenarios of tall buildings. The WTC7 fire (NIST, 2008) showed that fires in open floorplan offices travel through large compartments generating both areas of intense localized heating and of slow pre-heating, as well as areas of cooling. These occur simultaneously within the floor, thus producing both long-cool fires and short-hot fires (Lamont et al., 2004) as well as asymmetries with differential thermal expansion. The effects of these heterogeneities should be taken into consideration when translating temperature into heat fluxes to define the thermal loading in tall building design. Therefore, when designing a tall building in fire, the whole building behavior needs to be analyzed to understand truly how a system will perform under fire loading.

3.5 FIRE SEVERITY

Fire severity is a way to determine the destructive impact of a fire, the temperatures that could cause failure of the structure. From a series of compartment tests, Ingberg (1928) suggested that fire severity of a real fire could be calculated by considering equivalence of the areas under the standard fire temperature curve and the real fire temperature curve in an compartment above a base of either 150°C or 300°C as shown in Figure 3.6. It allows engineers to assess the severity of a fire in compartment based on the standard fire time–temperature curve. This method takes several factors into consideration such as ventilation, fire loads, and compartment size. It is a correlation between fire load measured in tests as load per unit floor area and the standard fire resistance periods.

Therefore, fire severity is most often defined in terms of the period of exposure to the standard test fire. The design of fire severity can be determined based on a complete burnout fire or the equivalent time of a complete burnout fire. The equivalent fire severity is the time of exposure to the standard fire that would result in the load-bearing capacity being the minimum occurring in a complete burnout of the fire cell. Law (1971) used a time equivalent method to define the equivalent fire severity using the time of exposure to the standard fire, which would result in the same maximum temperature in a protected steel member as would occur in a complete burnout of the fire test, as shown in Figure 3.7.

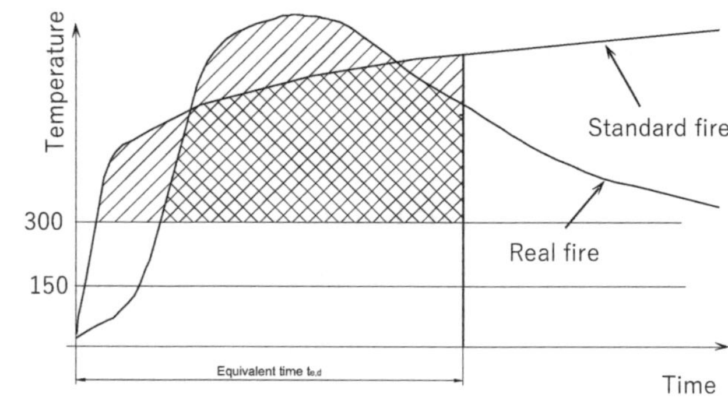

Figure 3.6 Method of Ingberg (1928) to determine fire severity.

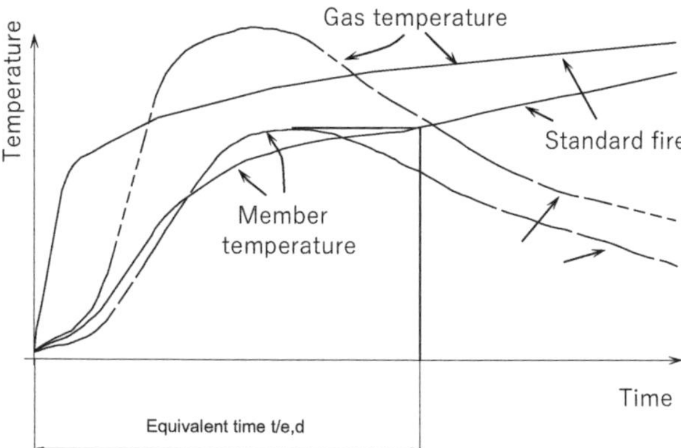

Figure 3.7 Equivalent time method to determine fire severity Law (1971).

3.6 FIRE LOAD

Fire loading is a measurement to determine the potential severity of a fire in a given space. It describes the amount of combustible material in a building or confined space and the amount of heat this can generate. The more the flammable materials in a space, the higher the fire load and therefore the faster the fire spreads. It is the heat output per unit floor area, often in kJ/m^2, calculated from the calorific value of the materials present.

A room with no furnishings and concrete walls will have a fire loading of nearly zero. If the fire spreads into such a room from elsewhere, it will find nothing to feed on. However, the presence of anything that is flammable such as furniture, electrical appliances or computer equipment, wood panels, acoustic tile, carpeting, curtains, or wall decorations will increase the fire loading. Buildings under construction or renovation tend to carry high fire loads in the form of construction materials, solvents, and fuel for generators.

3.6.1 Fire load calculation from Eurocode 1

Eurocode 1, EN 1991-1-2 (2002) provides the formula to calculate the fire load. Design value may be determined either from a national fire load classification of occupancies or by performing a fire load survey.

The design value of the fire load $q_{f,d}$ is defined as follows:

$$q_{f,d} = q_{f,k} \times m \times \delta_{q1} \times \delta_{q2} \times \delta_n \left[\text{MJ/m}^2\right] \tag{3.5}$$

where

m is the combustion factor,

δ_{q1} is a factor taking into account the fire activation risk due to the size of the compartment (see Table E.1 of Eurocode 1, EN 1991-1-2, 2002),

δ_{q2} is a factor taking into account the fire activation risk due to the type of occupancy (see Table E.1 of Eurocode 1, EN 1991-1-2, 2002),

δ_n is a factor taking into account the different active firefighting measures (sprinkler, detection, automatic alarm transmission, firemen),

$\delta_{f,k}$ is the characteristic fire load density per unit floor area [MJ/m^2].

3.6.2 Fire load density from Eurocode 1

In determination of fire load, factors such occupancy and floor area are used to characterize fire load densities $q_{f,k}\left[\text{MJ/m}^2\right]$, as given in Table 3.1, which shows fire load densities $q_{f,k}\left[\text{MJ/m}^2\right]$ for different occupancies.

Table 3.1 Fire load densities $q_{f,k}\left[\mathrm{MJ/m^2}\right]$ for different occupancies

Occupancy	*Average*	*80% Fractile*
Dwelling	780	948
Hospital (room)	230	280
Hotel (room)	310	377
Library	1,500	1,824
Office	420	511
Classroom of a school	285	347
Shopping center	600	730
Theater (cinema)	300	365
Transport (public space)	100	122

Note: Gumbel distribution is assumed for the 80% fractile.

Source: Reproduced from Table E.4 of EN 1991-1-2 (2002), Eurocode 1. Actions on structures, Part 1-2: General actions. Actions on structures exposed to fire. Commission of the European communities 2002; permission to reproduce and extracts from British and ISO standards is granted by BSI. British Standards can be obtained in PDF or hard copy formats from the BSI online shop: www.bsigroup.com/Shop or by contacting BSI Customer Services for hardcopies only: Tel: +44 (0)20 8996 9001, Email: cservices@bsigroup.com.

3.7 FIRE SPREAD

In tall buildings, the fire spreads in different ways. The direction and the speed of its spread are determined primarily by the compartment of the buildings. The potential for the spread of fire from an enclosure will be influenced by the thermal and mechanical responses of the enclosure's boundaries (walls, roof, doors, windows). The thermal and mechanical responses of boundary elements may be evaluated individually, subject to checking for interaction effects between adjacent structural members. For example, the thermal bowing of walls may affect the support or loading capacity of the enclosure's roof.

3.8 ROUTES OF FIRE SPREAD

Once the compartment is characterized, the designer should identify all the possible routes of fire transmission through the boundary surfaces. Figure 3.8 (PD 7974-3, 2019) illustrates some of the most common routes of potential fire spread. These routes of fire spread should be examined as a series of direct spread mechanisms. Ideally, all the potential routes of fire spread from the enclosure should be investigated and the minimum time for fire spread determined. However, design effort may be reduced in situations where expert judgment can identify the routes that are susceptible to the most rapid fire spread. It should be remembered that the determination as to whether or not fire spread takes place will be influenced by conditions both within the fire enclosure and within the adjacent enclosures.

3.8.1 Horizontal spread of fire

As shown in Figure 3.8, horizontal spread of fire is primarily through wall (if it has relatively low fire resistance), opening in the wall, ceiling, and void or duct in ceiling or floor.

3.8.2 Vertical spread of fire

As shown in Figure 3.7, vertical spread of fire can be primarily divided into two categories: internal and external.

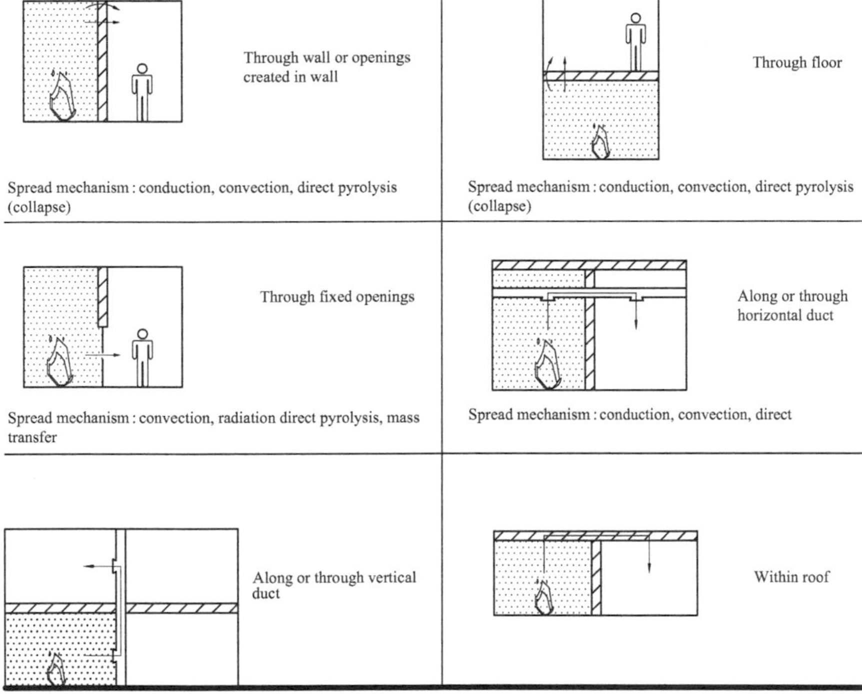

Figure 3.8 Routes of fire transmission. (Reproduced from Figure 4 of BSI "PD7974-3 (2019), Application of fire safety engineering principles to the design of buildings, Part 3: Structural response and fire spread beyond the enclosure of origin (Sub-system 3)," The British Standards Institute; permission to reproduce and extracts from British and ISO standards is granted by BSI. British Standards can be obtained in PDF or hard copy formats from the BSI online shop: www.bsigroup.com/Shop or by contacting BSI Customer Services for hardcopies only: Tel: +44 (0)20 8996 9001, Email: cservices@bsigroup.com.)

(Continued)

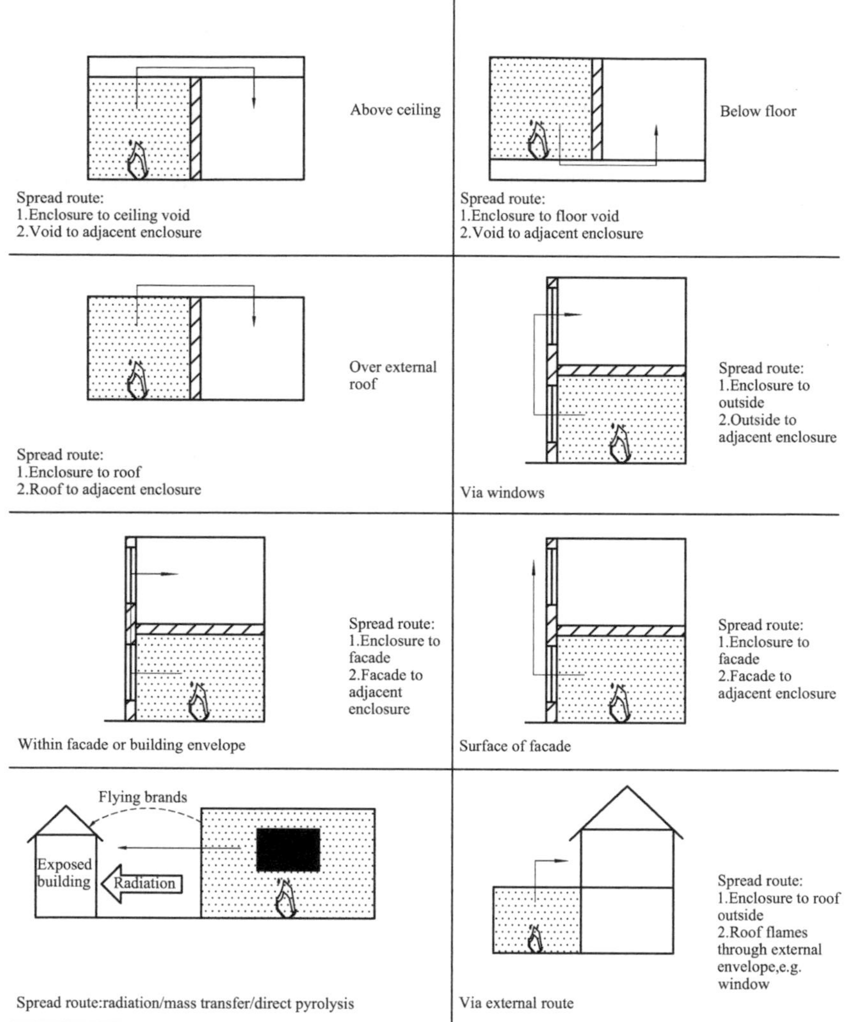

Figure 3.8 (Continued) Routes of fire transmission. (Reproduced from Figure 4 of BSI "PD7974-3 (2019), Application of fire safety engineering principles to the design of buildings, Part 3: Structural response and fire spread beyond the enclosure of origin (Sub-system 3)," The British Standards Institute; permission to reproduce and extracts from British and ISO standards is granted by BSI. British Standards can be obtained in PDF or hard copy formats from the BSI online shop: www.bsigroup.com/Shop or by contacting BSI Customer Services for hardcopies only: Tel: +44 (0)20 8996 9001, Email: cservices@bsigroup.com.)

3.8.2.1 *Fire spread through ducts, shafts, and penetrations (internal)*

For the 19-floor Windsor Tower in Spain (Fletcher, 2006), the fire spread to most parts of the building within 7h. The fire spread internally vertically through ducts, shafts, penetrations, etc. A fire of this nature will generally propagate extremely quickly without any hope of being controlled by sprinklers and has the potential of almost simultaneously compromising the life of everyone remaining within the building.

3.8.2.2 *Fire spread through façade*

As introduced in Fu (2017), in Grenfell Tower fire, the fire spread primarily from the façade. The fire originated inside a room can also spread through the window to the façade and subsequently spread to other floors. The fire can spread either through the exterior of the façade or the interior gaps. To avoid the first route of spread, the material used in the façade should have sufficient fire resistance. As shown in Chapter 2, if the façade is flammable, it will accelerate the spread of fire. To avoid the second route of spread, sufficient fire blocks should be designed to stop the spread of fire through gaps.

Apart from Grenfell Tower, vertical spread of fire through the façade has also been noticed in other fire incidents. In CCTV tower fire in China, the fire spread to most parts of the building in around 15min. It spread predominantly through cladding following an ignition in the cladding from a firework.

3.9 STRUCTURAL FIRE DESIGN

As mentioned in Chapter 2, structural fire design is one of the key fire safety design tasks. The basic principles will be introduced in this section, and more specific design methods will be introduced in detail in Chapter 4.

The objective of structural fire design is to determine the thermal behavior of the structural members and finding effective means for fire protections to satisfy fire safety design objectives.

Structural fire design generally consists of the following:

- Assessment of thermal response and structural response for different types of building members;
- Design structural system and its various components including supports and joints;
- Choice of size of structural members and fire protections with specified thermal and mechanical properties.

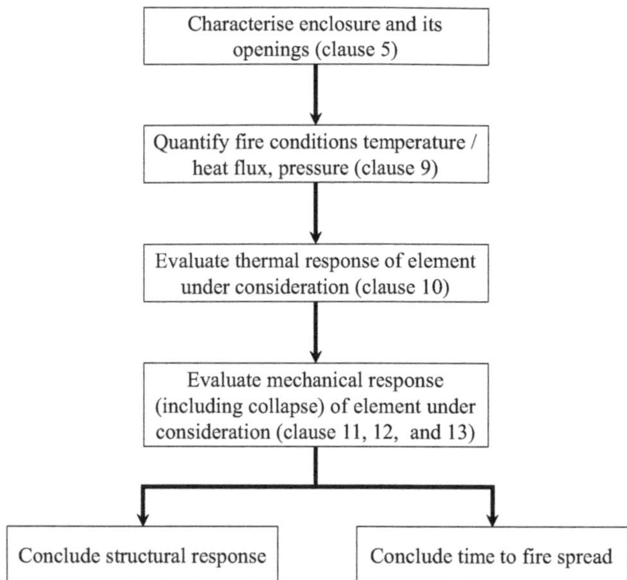

Figure 3.9 (Figure 12 of BSI "PD7974-3 (2019), Application of fire safety engineering principles to the design of buildings, Part 3: Structural response and fire spread beyond the enclosure of origin (Sub-system 3)," The British Standards Institute.) (Permission to reproduce and extracts from British and ISO standards is granted by BSI. British Standards can be obtained in PDF or hard copy formats from the BSI online shop: www.bsigroup.com/Shop or by contacting BSI Customer Services for hardcopies only: Tel: +44 (0)20 8996 9001, Email: cservices@bsigroup.com)

PD7974-3 (2019) specifies the procedure for designing structural members to avoid fire spread. The first step "characterize the enclosure and its opening" has been already introduced in Section 3.4.1. The remaining steps will be introduced in the following sections (Figure 3.9).

3.9.1 Determine the compartment temperature (design fire)

The first task in structural fire design is to determine design fire based on fire scenarios. Normally the design fire should be applied in one fire compartment of the building at a time, unless otherwise specified in the design fire scenario.

As introduced in Section 3.3, the simplest way to predict the time–temperature curve for a fire compartment is a standard fire temperature model in Eurocode. Parametric fire curve is closer to a real fire in predicting the heating rate and the maximum temperature of the atmosphere inside the fire compartment. Engineers need to assess the fire load (the quantity and type of combustible material), the ventilation, and the thermal characteristics of the compartment linings. These variables can be used to calculate the fire temperature per Eurocode. Using engineering software such as Zone model and computational fluid dynamics is a more accurate way.

3.9.2 Determine the thermal response of structural members

During the process of fire, heat will be transferred to the building elements through the so-called heat transfer process. The thermal response of each structural member can be worked out based on the basic thermodynamics' principles. Heat transfer is comprised of three processes: conduction is the mechanism of heat transfer in solid materials, in the steady-state situation; convection is the heat transfer by the movement of fluids, either gases or liquids; radiation is the transfer of energy by electromagnetic waves.

The increase in temperature for both internal unprotected or protected steelwork and concrete structural elements can be obtained from the fire tests given in BS 476: Parts 20–22 (1987).

Eurocode 3 (2005) and Eurocode 4 (2005) give the formulas to determine the increase in temperature in both protected and unprotected structural members. These formulas are based on the principles of heat transfer which will be briefly introduced here.

3.9.3 Heat transfer

3.9.3.1 Thermodynamics of heat transfer

Thermodynamics is the study of macroscopic continuum. Following energy conservation rules, a system of fixed mass must remain at a constant total energy if it is isolated from its surroundings. Under fire, only internal energy (U) will be significantly changed (excluding kinetic, electrical, etc.).

From the first law of thermodynamics, under fire we can get:

$$U_2 - U_1 = Q + W \tag{3.6}$$

where

U is the internal energy,
Q is the heat,
W is the work.

The rate of change in the internal energy will be

$$\frac{dU}{dt} = \dot{Q} + \dot{W} \tag{3.7}$$

In heat transfer, heat can flow in two distinct mechanisms due to the temperature difference as per the Fourier law:

1. **Conduction equation**

$$q = -\lambda \nabla \theta \tag{3.8}$$

where
q is the vector of heat flux per unit area,
λ is the thermal conductivity,
θ is the temperature.
The conservation of energy with the Fourier law requires

$$P_c\left(\frac{\partial \theta}{\partial t}\right) = -\nabla(\lambda \nabla \theta) + Q \tag{3.9}$$

where
ρ is the density,
c is the specific heat,
t is the time,
Q is the internal heat generation rate per unit volume.
Appropriate boundary conditions and initial conditions are needed to solve this equation.

2. **Convection and radiation**

Heat transfer to the surface of the structural member in a fire involves both convection and radiation. The net heat flux to the surface of the structural member can be expressed as follows:

$$Q_{\text{net}} = Q_{\text{net},c} + Q_{\text{net},r} \tag{3.10}$$

where
$Q_{\text{net},c}$ is the net convective heat flux per unit surface $\dot{h}_{\text{net}}, c = \alpha_c(\theta_g - \theta_m)$,
$\dot{Q}_{\text{net},r}$ is the net radiative heat flux per unit surface $h_{\text{net,r}} = \_\varepsilon_m \varepsilon_f \sigma \left[(\theta_g + 273)_4 - (\theta_m + 273)_4\right]$,
where
α_c is the convection heat flux coefficient,
θ_g is the gas temperature,
θ_m is the surface temperature of the structural member,
m is the surface emissivity of the structural member,
ε_f is the emissivity of the fire,
$\sigma = 5.67 \times 10^{-8}$ W/m^{2}K^4 is the Stefan–Boltzmann constant.

3.9.3.2 Eurocode formula to determine member temperature

3.9.3.2.1 Unprotected steel Section

For unprotected steel section, the increase in temperature within a small time interval is given by BS EN 1993-1-2: Eurocode 3 (BSI, 2005) and BS EN 1994-1-2: Eurocode 4 (BSI, 2005) as follows:

$$\Delta\theta_{a,t} = k_{\text{sh}} \frac{A_m / V}{c_a \rho_a} \dot{h}_{\text{net}} \Delta t \tag{3.11}$$

where

$\Delta\theta_{a,t}$ is the increase of temperature,
$A_{m/}V$ is the section factor for unprotected steel member,
A_m is the exposed surface area of the member per unit length,
V is the volume of the member per unit length,
c_a is the specific heat of steel,
ρ_a is density of the steel,
$\dot{h}_{\text{net}}$ is the designed value of the net heat flux per unit area,
Δt is the time interval,
k_{sh} is the correction factor for the shadow effect.

3.9.3.2.2.1 PROTECTED STEEL SECTION

For protected steel section, the increase in temperature within a small time interval is given by BS EN 1993-1-2: Eurocode 3 (BSI, 2005) and BS EN 1994-1-2: Eurocode 4 (BSI, 2005) as follows:

$$\Delta\theta_{a,t} = \left\{ \frac{\lambda_p / d_p}{c_a \rho_a} \frac{A_p}{V} \left(\frac{1}{1 + \Phi/3} \right) \left(\theta_{g,t} - \theta_{a,t} \right) \Delta t \right\} - \left\{ \exp(\Phi/10) - 1 \right\} \Delta\theta_{g,t} \tag{3.12}$$

where

$$\Phi = \frac{c_p \rho_p}{c_a \rho_a} d_p A_p / V$$

$\theta_{a,t}$ is the temperature of the steel at time t,
$\Delta\theta_{a,t}$ is the increase in temperature,
$\Delta\theta_{g,t}$ is the gas temperature at time t,
$\Delta\theta_{g,t}$ is the increase in the gas temperature,
$A_{p/}V$ is the section factor for protected steel member,
c_a is the specific heat of steel,

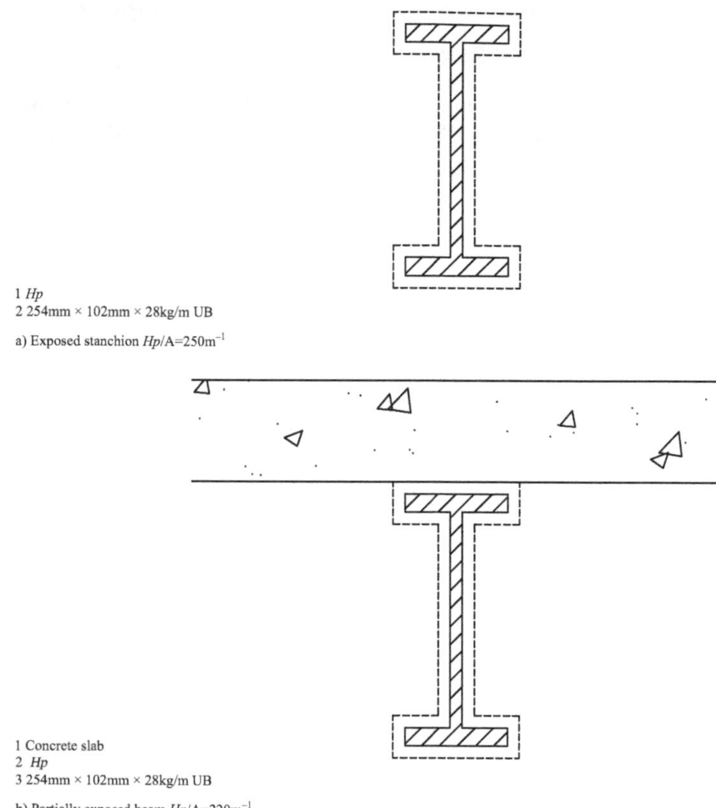

Figure 3.10 Calculation of section factor. (Reproduced from Figure 12 of BSI "PD7974-3 (2019), Application of fire safety engineering principles to the design of buildings, Part 3: Structural response and fire spread beyond the enclosure of origin (Sub-system 3)," The British Standards Institute; permission to reproduce and extracts from British and ISO standards is granted by BSI. British Standards can be obtained in PDF or hard copy formats from the BSI online shop: www.bsigroup.com/Shop or by contacting BSI Customer Services for hardcopies only: Tel: +44 (0)20 8996 9001, Email: cservices@bsigroup.com.)

c_p is the specific heat of fire protection material,
ρ_a is the density of the steel,
ρ_p is the density of the fire protection material,
d_p is the thickness of the fire protection material,
λ_p is the thermal conductivity of the fire protection material,
Δt is the time interval.

As shown in Figure 3.11, under the same standard fire temperature, there is slight temperature difference between protected and unprotected structural steel members.

3.9.3.2.1.2 SECTION FACTOR

As shown in Figure 3.10, the section factor is one of the important factors in calculating the temperature of the steel members. It is defined based on both the geometry and configuration of the members used in a building. For unprotected members, the heated perimeter A_m depends on the type of insulation. For protected members, the heated perimeter Am depends on the type of insulation, for example, sprayed insulation or intumescent paint which is applied to the section profile or board insulation which boxes the section and on the number of sides of the members exposed to the effect of the fire.

3.9.4 Material degradation at elevated temperatures

The strength for the Material properties of steel and concrete start to lose at elevated temperatures. Eurocode gives the material degradation for both concrete and steel material, which will be introduced here.

3.9.4.1 Degradation of steel material in fire

Figure 3.12 shows the stress–strain curves of steel at different temperatures from EN1994-1-2 Eurocode 4 (2005). The loss of strength can be illustrated by the amount of the stress that the member is able to withstand before reaching 2% strain.

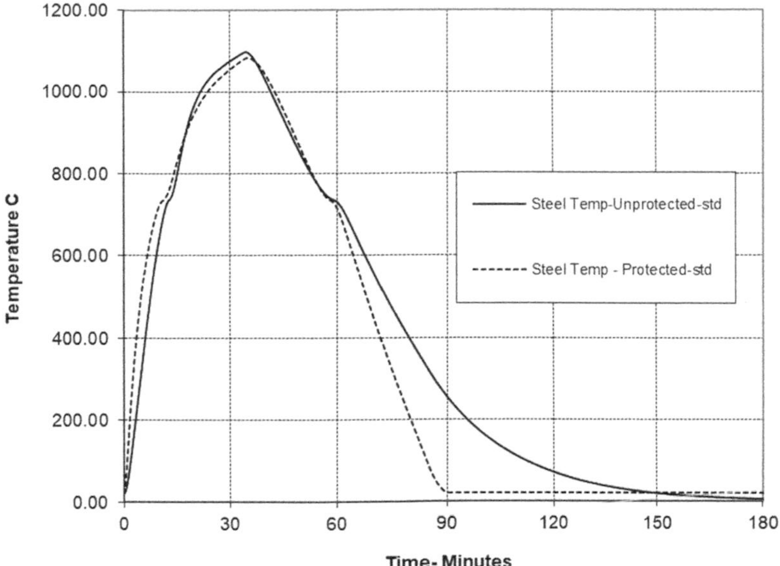

Figure 3.11 Temperatures of steel members (protected and unprotected members).

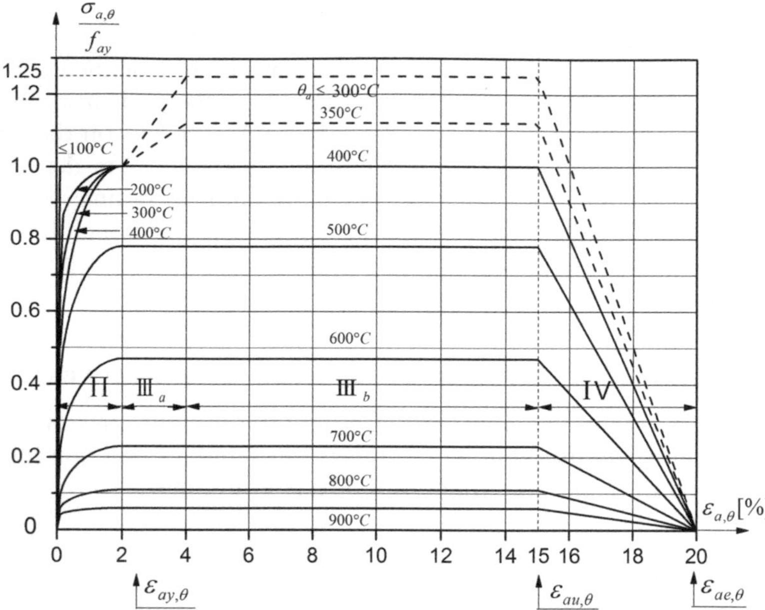

Figure 3.12 Stress–strain relationships of structural steel at elevated temperatures, strain-hardening included.(Figure A.1 of EN1994-1-2 (2005), incorporating corrigendum July 2008, Eurocode 4 Design of composite steel and concrete structures, Part 1-2: General rules. Structural fire design; permission to reproduce and extracts from British and ISO standards is granted by BSI. British Standards can be obtained in PDF or hard copy formats from the BSI online shop: www.bsigroup.com/Shop or by contacting BSI Customer Services for hardcopies only: Tel: +44 (0)20 8996 9001, Email: cservices@bsigroup.com.)

3.9.4.2 Degradation of concrete material in fire

Figure 3.13 is the stress–strain relationship of concrete from EN1994-1-2, Eurocode 4 (2005). The temperature shows a significant effect on the stress–strain relationships of concrete.

3.9.5 Design values of material properties under fire

In structural fire analysis, an Engineer should consider the material degradation mentioned in the proceeding sections. The design values of materials' mechanical properties in fire $X_{d,fi}$ are defined in Formula 2.1 of Eurocode 2 (EN1992-1-2, 2004) as follows:

$$X_{d,fi} = \frac{k_\theta X_k}{\gamma_{M,fi}} \tag{3.13}$$

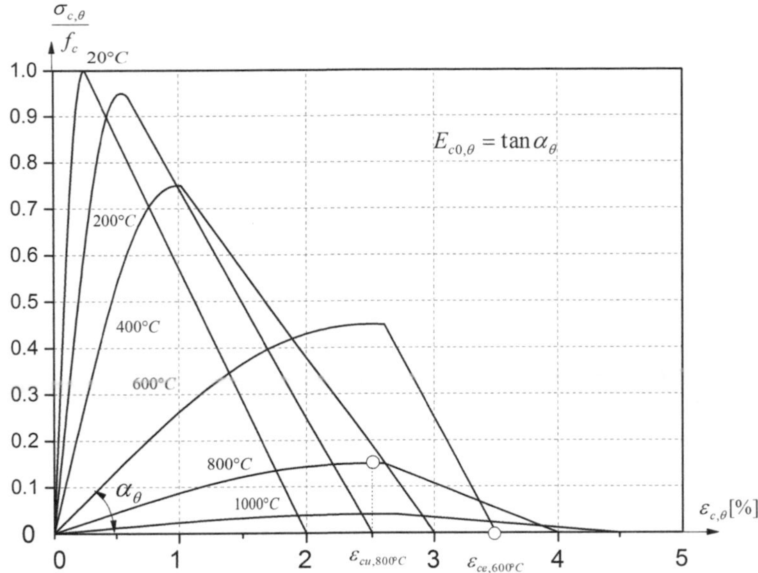

Figure 3.13 Behavior of compressive concrete material. (Figure B.1 of EN1994-1-2 (2005), incorporating corrigendum July 2008, Eurocode 4 Design of composite steel and concrete structures, Part 1-2: General rules. Structural fire design; permission to reproduce and extracts from British and ISO standards is granted by BSI. British Standards can be obtained in PDF or hard copy formats from the BSI online shop: www.bsigroup.com/Shop or by contacting BSI Customer Services for hardcopies only: Tel: +44 (0)20 8996 9001, Email: cservices@bsigroup.com.)

where

X_k is the characteristic value of a strength or deformation property for normal temperature design to EN 1992-1-1.

k_θ is the reduction factor for a strength or deformation property ($X_{k,\theta}/X_k$) dependent on the material temperature (see Figure 3.14). It shows the reduction factor for structural steel members defined by EN1994-1-2 Eurocode 4 (2005).

$\gamma_{M,fi}$ is the partial safety factor for the relevant material property in the fire situation.

3.9.6 Design of structural members in fire

3.9.6.1 *Mechanical design approaches of structural members in fire*

Eurocode 1 (EN 1991-1-2, 2002) specifies the requirements for mechanical design of structural members in fire. It should start with a temperature analysis; when performing the temperature analysis of a member, the

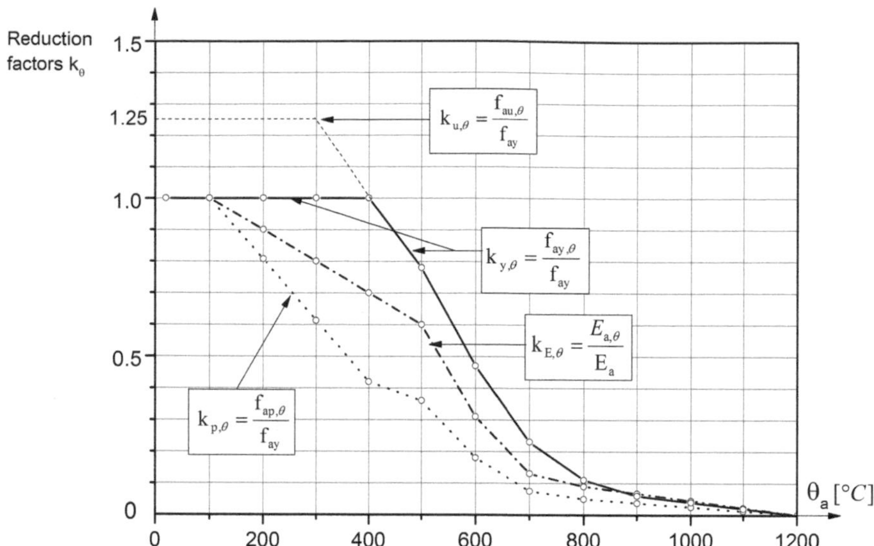

Figure 3.14 Reduction factors for stress–strain relationships allowing for strain-hardening of structural steel at elevated temperatures. (Figure A.2 of EN1994-1-2 (2005), incorporating corrigendum July 2008, Eurocode 4 Design of composite steel and concrete structures, Part 1-2: General rules. Structural fire design; permission to reproduce and extracts from British and ISO standards is granted by BSI. British Standards can be obtained in PDF or hard copy formats from the BSI online shop: www.bsigroup.com/Shop or by contacting BSI Customer Services for hardcopies only: Tel: +44 (0)20 8996 9001, Email: cservices@bsigroup.com.)

position of the design fire in relation to the member should be taken into account. Either standard fire or parametric fire temperature can be used in the analysis. The difference between them is whether the temperature analysis of the structural members is made for the full duration (including the cooling phase) or not.

After the temperature analysis, the mechanical analysis should be performed for the same duration as in the temperature analysis. The analysis can be performed in three domains.

1. In the time domain:

$$t_{fi,d} > t_{fi,\text{requ}}$$

2. In the strength domain:

$$R_{fi,d,t} > E_{fi,d,t}$$

3. In the temperature domain:

$$\Theta_d \leq \Theta_{cr,d}$$

where

$t_{fi,d}$ is the design value of the fire resistance,

$t_{fi,requ}$ is the required fire resistance time,

$R_{fi,d,t}$ is the design value of the resistance of the member in the fire situation at time t,

$E_{fi,d,t}$ is the design value of the relevant effects of actions in the fire situation at time t,

Θ_d is the design value of material temperature,

$\Theta_{cr,d}$ is the design value of the critical material temperature.

The specific mechanical design approaches in both temperature domain and strength domain will be introduced in Chapter 4.

3.9.6.2 *The acceptance criteria in designing structural members for tall buildings*

For designing the structural element of a tall building under fire, an engineer can use several acceptance criteria as follows:

- The overall stability of the structure is maintained. This indicates no local collapse or global collapse will be triggered due to the failure or deflection of the structural member to be designed.
- Deflections of the slab or beams on any lines of compartmentation are within the allowance of the deflection of the compartment wall where relevant. This is for the integrity of the compartment to ensure that the compartment wall is still functioning during the fire.
- The maximum tensile plastic strain at the level of the reinforcement is less than 5%. This is conservative in comparison with normal elongation values of 13%–15% for steel reinforcement but allows for the smeared cracking effect.
- In addition, the relative deflections of the main beams will be checked to ensure that deflections do not exceed span/20 at the relevant fire resistance period required for that area of the structure. A similar check for the slab should also be made.

The acceptance criteria can be achieved through a detailed structural fire analysis. Some criteria may be considered satisfied where the minimum thickness of walls or slabs is specified in accordance with Table 5.3 of EN 1991-1-2 (2002).

3.10 FIRE RESISTANCE

As stated in BS 476-20 (1987), "Fire resistance is the time which an element of building construction is able to withstand exposure to a standard temperature and pressure regime without a loss of its fire separating function or load bearing function or both." Fire resistance is a measure of the ability of the structure to resist collapse and to prevent the spread of fire during exposure to a fire of specified severity.

3.10.1 Methods to determine fire resistance

The fire resistance of any individual element may be determined by

a. Standard fire resistance tests,
b. Experimental large- and small-scale fire tests,
c. Expert assessment,
d. Quantitative analysis of fire spread mechanisms.

3.10.2 Fire resistance rating

Fire resistance rating is the fire resistance assigned to a building element on the basis of a test or some other approval system. Some countries use other terms such as fire rating, fire endurance rating, or fire resistance level. These terms are usually interchangeable. Fire resistance ratings are most often assigned in whole numbers of hours or parts of hours, in order to allow easy comparison with the fire resistance requirements specified in building codes. For example, a wall that has been shown by test to have a fire resistance of 75 min will usually be assigned a fire resistance rating of 1 h.

3.10.3 Fire resistance test for load-bearing structural members

Each building element needs to be assigned a fire resistance rating for comparison with the fire severity specified by codes. The most common method for assessing fire resistance is to carry out a full-scale fire resistance test. As shown in Figures 3.15 and 3.16, the building members can be placed into a fire furnace to heat up. In the meantime, it is loaded to

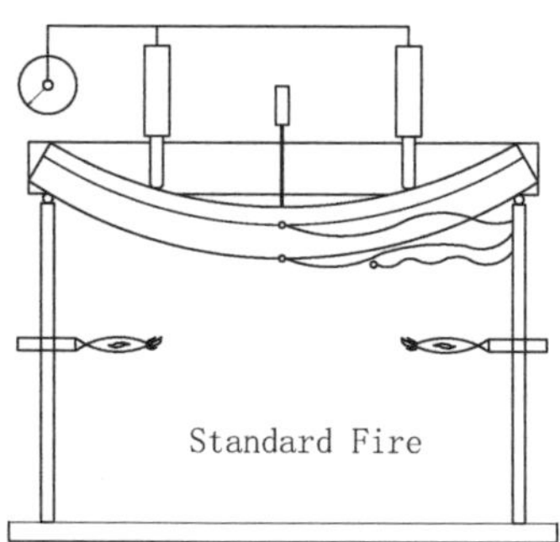

Figure 3.15 PD 7974-3:2003 Standard fire resistance tests.

Figure 3.16 Fire test furnace. (Photo taken by the author.)

failure. The time to failure can therefore be determined. For fire resistance testing, many countries use the ISO 834-11 (2019) or have other national standards based on it.

3.10.4 Fire resistance requirements for elements of a tall building

An important lesson learned from the WTC failures (NIST, 2008) and Cardington fire tests is that the prescriptive fire resistance ratings of individual building elements do not guarantee that a building system as a whole will perform adequately. A holistic performance-based fire design approach is still needed for tall buildings, which will be introduced in Chapter 5.

In tall buildings, prescriptive fire resistance ratings of individual building elements do not guarantee a building system that as a whole will perform adequately. Therefore, fire resistance requirements should be based on the design fire scenarios, and the factors need to be considered are probability of fire occurrence, fire spread, fire duration, fire load, severity of fire, ventilation, compartment (type, size, geometry), and type of structural elements. In addition, the height of the building, number of occupants and type of their activities, sprinklers, and other active firefighting measures should also be considered.

3.11 FIRE PROTECTION METHOD

A fire protection system offers a fire-resisting period—the time needed to evacuate the occupants of a building—that varies between 30, 60, 90 or 120 min. There are two main categories of fire protection methods: one is active control, and the other is passive control.

3.11.1 Active control system

Active control system provides fire protections through the actions taken by a person or an automatic device such as sprinkler or firefighters, control smoke (Figure 3.17).

3.11.2 Passive control system

The fire protection systems are built into the structural elements of a building such as intumescent paint, spray, and board protection on the structural steel members.

3.11.2.1 Intumescent paints

An intumescent coating is basically made a paint-like material which is inert at low temperatures—but reacts with heat at high temperatures. As the temperature rises during a fire event, the intumescent coating swells

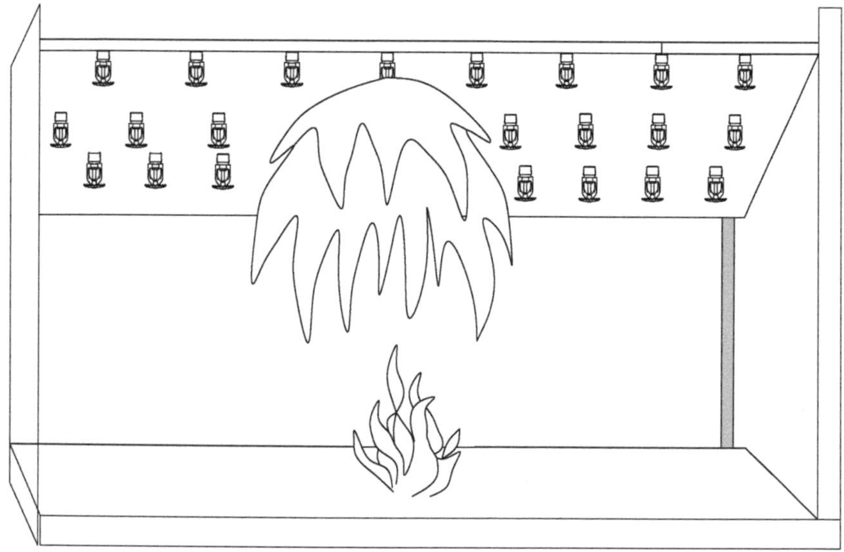

Figure 3.17 A typical sprinkler system in a room.

and forms a char layer that covers the steel. This char layer is of low thermal conductivity, thus acting as an insulating system. It should be noted that the coating usually expands up to 50 times when compared to its original thickness. The pros and cons of this type of protection are as follows:

Advantages
- Decorative finish
- Rapid application
- Easy-to-cover complex details
- Easy post-protection
- Fixings to steelwork, for example, service duct hangers
- Quicker construction
- Improved quality control
- Reduction in site disruption
- Cleaner sites
- Improved site safety
- Easier servicing installation.

Limitations
- Suitable for dry internal environments only
- More expensive than spray
- Require cleaned surfaces.

3.11.2.2 Spray fire protection

In this method, a certain fire-resistive material is sprayed on the structural members on site. The appropriate thickness of the spray-applied fire-resistive material is determined by a standard fire testing.

Advantage
- Fire-protective insulation can be applied by spraying to almost any type of steel member
- Most products can achieve up to 4 h of protection
- Low cost
- Rapid application
- Easy-to-cover complex details
- Often applied to non-primed steelwork.

Disadvantages
- Appearance is poor for visible members
- Overspray may need masking or shielding
- Primer, if used, must be compatible.

3.11.2.3 Board fire protection

This method uses either gypsum board wrappings or insulation board enclosures on the structural members. The pros and cons of this type of protection are as follows:

Advantages

- Clean, dry fixing
- Boxed appearance suitable for visible members
- Guaranteed thickness.

Disadvantages

- Require careful fitting around complex detail
- More expensive than sprayed protection.

3.11.3 Fire resistance test for protected members

Similar to the fire test of unprotected member, the protected members can be placed into a fire furnace to heat up. The test procedure can follow ISO 834-11 (2014). In the meantime, it is loaded to failure. The time to failure can therefore be determined. Standard fire temperature will be used as well in the test.

Two types of tests will be performed:

Insulation tests

These tests determine the thickness of fire protection needed to keep the average steel temperature at or below 550°C after a given fire resistance period.

Stickability tests

These tests ensure that the fire protection remains intact over the fire resistance period.

REFERENCES

BS 476-20 (1987), Incorporating Amendment No. 1. Fire tests on building materials and structures, Part 20: Method for determination of the fire resistance of elements of construction (general principles).

BSI 'PD7974-3 (2019), Application of fire safety engineering principles to the design of buildings, Part 3: Structural response and fire spread beyond the enclosure of origin (Sub-system 3)', The British Standards Institute.

Ellobody, E., Bailey, C. G. (2011, June), Structural performance of a post-tensioned concrete floor during horizontally travelling fires. *Engineering Structures*, 33(6), pp. 1908–1917.

EN 1991-1-2 (2002), Eurocode 1. Actions on structures, Part 1-2: General actions. Actions on structures exposed to fire. Commission of the European communities.

EN 1992-1-2 (2004), Eurocode 2. Design of concrete structures, Part 1-2: General rules. Structural fire design. Commission of the European communities.

EN 1993-1-2 (2005), Eurocode 3. Design of steel structures, Part 1-2: General rules. Structural fire design. Commission of the European communities.

EN 1994-1-2 (2005), Eurocode 4. Design of composite steel and concrete structures, Part 1-2: General rules. Structural fire design. Commission of the European communities.

Fletcher, I., AHitchen, B., Welch, S. (2006), Performance of concrete in fire: A review of the state of the art, with a case study of the Windsor Tower fire, in: *Proceedings of the 4th International Workshop in Structures in Fire*, pp. 779–790.

Fu, F. (2017). Grenfell Tower disaster: how did the fire spread so quickly? *The Conversation*.

Horová, K., Tomáš, J., František, F. (2013, August–September), Temperature heterogeneity during travelling fire on experimental building. *Advances in Engineering Software*, 62–63, pp. 119–130.

Ingberg, S. H. (1928), Tests of the severity of building fires. *National Fire Protection Association*, 22, pp. 43–61.

ISO 834-11 (2014), Fire resistance tests—Elements of building construction, Part 11: Specific requirements for the assessment of fire protection to structural steel elements, ISO/TC 92/SC 2 Fire containment.

ISO 834-2 (2019), Fire-resistance tests—Elements of building construction, Part 2: Requirements and recommendations for measuring furnace exposure on test samples, ISO/TC 92/SC 2 Fire containment.

Lamont, S., Usmani, A. S., Gillie, M. (2004), Behaviour of a small composite steel frame structure in a "long-cool" and a "short-hot" fire. *Fire Safety Journal*, 39(5), p. 327.

Law, M. A (1971), Relationship between fire grading and building design and contents. Fire Research Note Number 877. Fire Research Station, UK.

NIST NCSTAR (2005, December), Federal building and fire safety investigation of the World Trade Center disaster, final report of the National Construction Safety Team on the collapses of the World Trade Center Towers.

NIST NCSTAR 1A (2008), Final report on the collapse of world trade center building 7, National Institute of Standards and Technology, US Department of Commerce.

Rein, G., Zhang, X., William, P., Hume, B., Heise, A., Jowsey Lane, A., Lane, B., Torero, J. L. (2007), Multi-storey fire analysis for high-rise buildings, in: *Proceedings of the 11th International Interflam Conference, London, UK*, pp. 605–616.

Zhang, C., Li, G. Q., Usmani, A. (2013), Simulating the behavior of restrained steel beams to flame impingement from localized-fires. *Journal of Constructional Steel Research*, 83, pp. 156–165.

Chapter 4

Structural fire design principles for tall buildings

4.1 INTRODUCTION

In this chapter, the structural fire design principles will be introduced in depth on the basis of Chapter 3. It introduces the structural fire design procedures for steel, concrete, and composite structural members based on Eurocodes and British Standards. The two key structural fire design methods, namely, critical temperature method and moment capacity method, are both introduced. It also covers the design of post-tensioning slabs, connection, and beams with openings.

4.2 KEY TASKS FOR STRUCTURAL FIRE DESIGN

As introduced in the preceding chapters, structural fire design is one of the key design tasks in fire safety design. This is because an effective structural fire design will assist in both evacuation route design and compartmentation design.

The primary task for structural fire design is to design the fire resistance of structural (load-bearing) members such as beams, columns, wall, and slabs to ensure that they won't fail within the required fire rating duration specified by various codes. In addition, the design shall ensure the structural stability of the whole building or local area of a building to avoid the entire or partial collapse of the building. Prevention of fire-induced collapse will be discussed in detail in Chapter 7.

4.2.1 Building elements to be considered in design for fire

In a tall building, building elements to be considered primarily in design for fire are as follows:

- Structural frames (including structural beams and columns and the frame as a whole)
- Floors (concrete floors or composite floors)
- Load-bearing walls (such as shear walls and core walls)
- Compartment walls (can be either load-bearing or non-load-bearing)

Not included: Roofs—unless they are also designed to be part of means of escape.

4.2.2 Design of structural members in fire

As introduced in Chapter 2, the structural members in fire can be designed in three domains: time domain, strength domain, or temperature domain. Two methods are recommended by the codes. One is in the temperature domain, called critical temperature method; the other is in the strength domain, called moment capacity method. They will be introduced in this chapter.

4.2.3 Design procedures

In Chapter 3, the structural fire design steps specified in PD7974-3(2019) are introduced. The key steps can be further simplified as shown in Figure 4.1.

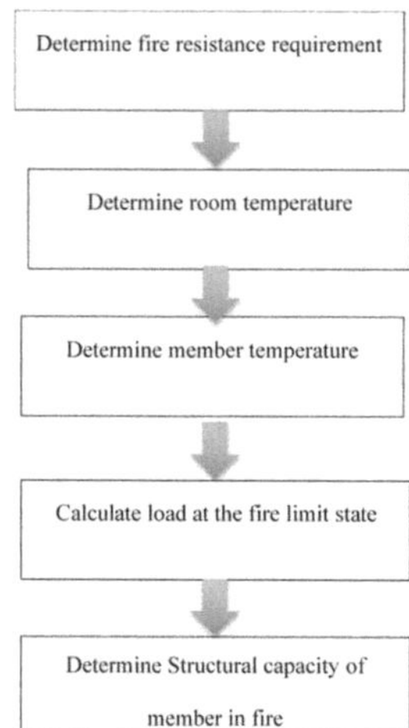

Figure 4.1 Key structural fire design procedure.

Some of the design steps has been provided in Chapter 3; therefore, the last two steps will be introduced in detail in this chapter.

4.3 FIRE RESISTANCE RATING FOR LOAD-BEARING STRUCTURAL MEMBERS

British Approved Document B (2019) Section 6, clause 6.1 specifies that "Elements such as structural frames, beams, columns, load-bearing walls (internal and external), floor structures and gallery structures should have, as a minimum, the fire resistance given in Appendix B, Table B3."

Fire resistance requirements for a structural member should be based on the parameters influencing fire growth and development. In addition, the type of structure and other design aspects such as evacuation and firefighting access also affect the fire resistance requirements of a structural member. The key factors that should be considered in the design include:

1. Factors affecting fire scenarios
   Probability of fire occurrence, fire spread, fire duration, fire load, severity of fire, ventilation conditions, fire compartment (type, size, geometry)
2. Type of the structural element
   Steel, concrete or composite, or timber
3. Factors affecting evacuation and firefighting

   - Evacuation conditions
   - Access for fire rescue teams

Table B3 of Approved Document B (2019) is reproduced here in Table 4.1. It specifies minimum fire resistance time for major structural elements in a building.

4.4 DESIGN OF CONCRETE MEMBERS IN FIRE

As already said, degradation starts when concrete is heated up. Significant deterioration will be noticed in all concrete members after the temperature rises up to 550°C, caused by the differential rate of thermal expansion of the constituent materials. The reinforcement responded differently from the plain concrete.

Table 4.1 Fire resistance requirement for building members

Part of building minimum provisions when tested to the relevant European standard (minutes)	*Minimum provisions when tested to the relevant European standard (minutes)*	*Alternative minimum provisions when tested to the relevant part of BS 476 (minutes)*			*Type of exposure*
		Load-bearing capacity	*Integrity*	*Insulation*	
1. Structural frame, beam, or column	R see Table B4	See Table B4	Not applicable	Not applicable	Exposed faces
2. Load-bearing wall (which is not also a wall described in any of the following items)	R see Table B4	See Table B4	Not applicable	Not applicable	Each side separately
3. Floors					
a. Between a shop and flat above	REI 60 or see Table B4 (whichever is greater)	60 min or see Table B4 (whichever is greater)	60 min or Table (whichever is greater)	60 min or see Table B4 (whichever is greater)	From underside
b. In upper storey of two-storey dwelling house (but not over garage or basement)	R 30 and REI 15	30 min	15 min	15 min	From underside
Any other floor—including compartment floors	REI see Table B4	See Table B4	See Table B4	See Table B4	From underside
4. Roofs					
a. Any part forming an escape route	REI 30	30 min	30 min	30 min	From underside
b. Any roof that performs the function of a floor	REI see Table B4	See Table B4	See Table B4	See Table B4	From underside
5. External walls					
a. Any part a maximum of 1,000 mm from any point on the relevant boundary	REI see Table B4	See Table B4	See Table B4	See Table B4	Each side separately

(*Continued*)

Table 4.1 (Continued) Fire resistance requirement for building members

Part of building minimum provisions when tested to the relevant European standard (minutes)	*Minimum provisions when tested to the relevant European standard (minutes)*	*Alternative minimum provisions when tested to the relevant part of BS 476 (minutes)*			*Type of exposure*
		Load-bearing capacity	*Integrity*	*Insulation*	
b. Any part a minimum of 1,000 mm from the relevant boundary	RE see Table B4 and REI 15	See Table B4	See Table B4	15 min	From inside the building
c. Any part beside an external escape route (Section 2 Diagram 2.7 of Approved Document B Volume 1 and Section 3, Diagram 3.4).	RE 30	30 min	30 min	No provision	From inside the building
6. Compartment walls separating either					
a. Flat from any other part of the building (see paragraph 7.1 of Approved Document B Volume 1)	REI 60 or see Table B4 (whichever is greater)	60 min or see Table B4 (whichever is less)	60 min or see Table B4 (whichever is less)	60 min or see Table B4 (whichever is less)	Each side separately
b. Occupancies	REI 60 or see Table B4 (whichever is less)	60 min or see Table B4 (whichever is less)	60 min or see Table B4 (whichever is less)	60 min or see Table B4 (whichever is less)	Each side separately
c. Seven compartment walls (other than in item 6 or item 10).	REI see Table B4	See Table B4	See Table B4	See Table B4	Each side separately
8. Protected shafts excluding any firefighting shaft					
a. Any glazing	E 30	Not applicable	30 min	No provision	Each side separately
b. Any other part between the shaft and a protected lobby/corridor	REI 30	30 min	30 min	30 min	Each side separately
c. Any part not described in (a) or (b) above	REI see Table B4	See Table B4	See Table B4	See Table B4	Each side separately

Reproduced from Appendix B, Table B3 "Specific provisions of the test for fire resistance of elements of structure" of British HM Government Approved Document B (2019) Fire safety, In public domian on British HM Government website).

4.4.1 Thermal response of concrete in fire

When exposed to fire, physical and chemical changes occur in concrete as follows:

- At 100°C–140°C, evaporation of water starts.
- At 300°C, the cement paste starts to shrink due to water evaporation. Aggregate starts to expand with spalling of concrete may be observed.
- At 400°C–600°C, the calcium hydroxide in cement paste decomposes into calcium oxide and water, a significant reduction in strength.
- At >550°C, the aggregate in the concrete start to decompose, a significant loss in concrete strength.

The research from OFFSHORE TECHNOLOGY REPORT2001/074 shows that

- Different curing method affects deterioration of reinforced concrete from around 60% to 85%.
- Deterioration increased linearly slightly with the increasing rate of heating for the plain concrete but decreased nonlinearly when the concrete was reinforced.
- There was a tendency for more severe deterioration for the plain concrete, but this was not statistically significant.
- There was a faster increase in deterioration from plain lightweight aggregate concrete to normal concrete reinforced concrete.

4.4.2 Spalling

During exposure to fire, heat and mass transfer happens in concrete actively. Under steep temperature gradient, the buildup of water pressure develops high local stresses, which may cause concrete spalling. The combination of thermal stress and the pore pressure due to the evaporation of the free water inside the concrete makes the spalling to happen. Zhukov (1975) developed a spalling model as shown in Figure 4.2; in Zhukov's model, the stresses developed due to combined mechanical and thermal loads. He considered that the stresses acting could be categorized as load-induced stresses, thermal stresses, and pore pressures.

4.4.2.1 Types of spalling

The two types of spalling are as follows:

Explosive spalling occurs in a very early stage of a fire which is likely to lead to loss of protective cover of the main reinforcement. Rapid rises in temperature, such as short-hot fire, will result in strength loss

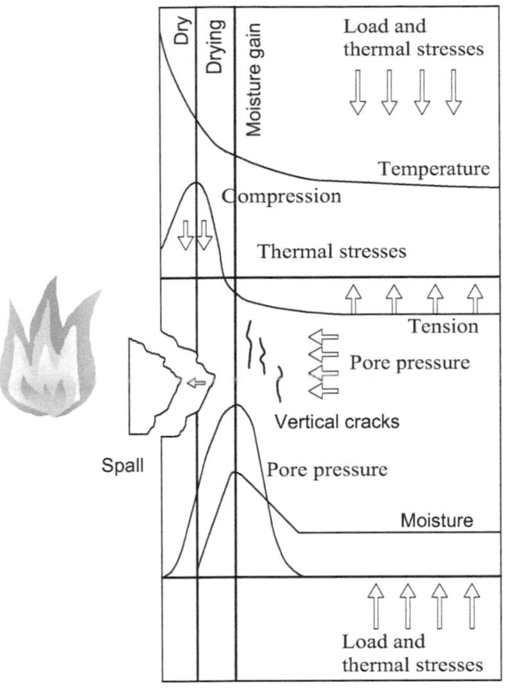

Figure 4.2 Zhukov's spalling model (Khoury and Anderberg, 2000).

leading to reduction in fire resistance. The key factors affecting explosive spalling are the rate of temperature rise, the restraint conditions to thermal expansion, and the permeability of the concrete.

Sloughing spalling, the concrete gradually comes away due to loss of effective bond and strength.

4.4.2.2 Prevention of spalling

The explosive spalling usually occurs during the first 7–30 min of a fire incident. The spalling is unlikely to occur when the moisture content of the concrete is less than 3% by weight. EC2 EN 1992-1-2:2004/A1:2019 (E) specifies that explosive spalling shall be avoided but its influence on performance requirements (such as EI) shall be taken into account. Several preventive measures that need to be taken to achieve this requirement are introduced here.

4.4.2.2.1 Thermal barrier

Thermal barriers usually limit the increase of temperature at the surface of the concrete. It includes fire-resistive materials—such as coating of fire-proof paints and plastering of fire-proof mortars—that form a protective barrier on concrete structural member, control temperature rise in concrete surface layer, and thus reduce temperature gradient and the risk of spalling.

4.4.2.2.2 Reduction of vapor pressure

The spalling happens primarily due to the evaporation pressure of the water; therefore, reducing the vapor pressure is an effective way of preventing the spalling. Several methods used for this are as follows:

- Addition of polypropylene (PP) fiber
- Forced drying of structural members
- Installation of moisture eliminatory tubes.

The most commonly used method in recent years is addition of PP fibers, which will be introduced here.

4.4.2.2.3 PP fiber

PP fiber (as shown in Figure 4.3) has been widely used recently in some of the projects (primarily for tunnels) to resist concrete spalling in fire. A large-scale fire test of European Concrete Building Project (ECBP) was conducted at BRE's Cardington in 2001. The flat slab was supported by a number of high-strength (C85) concrete columns containing 2.7 kg/m^3 PP fibers to reduce the tendency for explosive spalling. All of the columns survived the fire test without any significant spalling.

The addition of fibers in concrete matrix bridges cracks and restrains them from further opening. Figure 4.4 is the scanning electron microscopic

Figure 4.3 Polypropylene fiber.

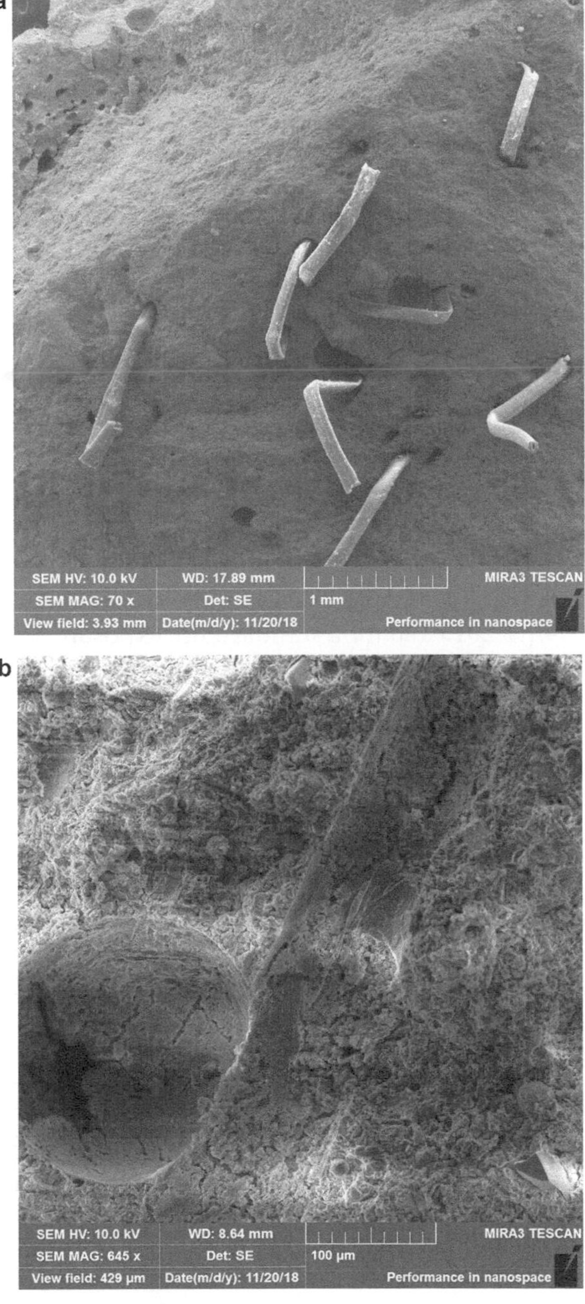

Figure 4.4 SEM images of fracture of different fiber in coral concrete. The behavior of PP fiber. a) PP fibre inside concrete before fire. b) Channels left by PP fibre (Cai et al., 2020).

image of the microstructure of PP fiber-mixed concrete under fire condition. It can be seen that PP fiber fills the void of the cement matrix. When temperature increases to 200°C, the PP fiber melts and leaves void channel (as shown in Figure 4.4). The channels can allow vapor to escape and thus reduce the buildup of high pore pressure within the concrete.

Research by Cai et al. (2020a) shows that for an exposure to a high temperature of 600°C for 6 h, the loss in compressive strength is about half of that when no PP fibers are used. The lower contents of fibers generally showed worse performance due to further deteriorations and higher volumes of voids. As the content of PP increases, the slump of the mix decreases. Thus, for heavily reinforced concrete members, it is recommended to use super-plasticizers to enhance the workability.

4.4.3 Simplified calculation methods for concrete members from EC2 EN 1992-1-2:2004/A1:2019 (E), 500°C isotherm method

EC2 EN 1992-1-2:2004/A1(2019) specifies a simplified method to design structural members under bending and axial loads in fire. This method is valid for minimum width of cross section given in Table B1 of EC2 EN 1992-1-2:2004/A1:2019 and for a standard fire exposure depending on the fire resistance, as well as for a parametric fire exposure with an opening factor (O) of ≥0.14 $m^{1/2}$.

This method is considering a reduction in the cross-sectional size of concrete beams due to heat damages on the concrete surfaces. The thickness of the damaged concrete is made equal to the average depth of the 500°C isotherm in the compression zone of the concrete section (Figures 4.5 and 4.6). This is a simplified method with simple assumption that concrete will lose all its bearing capacity when the member temperature rises to 500°C, while the residual concrete maintains its initial values of strength and elastic modulus.

On the basic of the above assumptions, after reaching the 500°C isotherm, a new width b_{fi} and a new effective height d_{fi} of the section will be calculated by excluding the concrete outside the 500°C isotherm. The rounded corners of isotherms can be regarded by approximating the real form of the isotherm to a rectangle or a square. Then determine the reduced strength of the reinforcement according to the temperature of reinforcing bars in the tension and compression zones. Finally, the bending capacity of the beam post fire can be obtained based on the reduction factor of materials and the formula of bending capacity. Conventional calculation methods can then be used to determine the load-bearing capacity of the reduced cross-section.

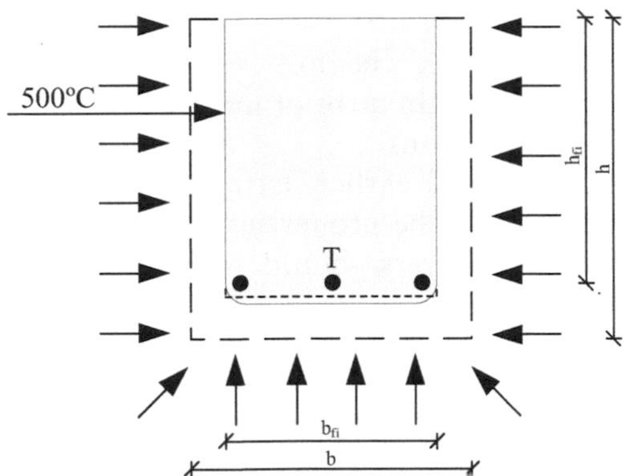

Figure 4.5 Reduced cross section of reinforced concrete beam and column with three-side fire exposure. (Reproduced from Figure B.1 of Page 72 BS EN 1992-1-2:2004+A1 (2019), Eurocode 2. Design of concrete structures, Parts 1–2: General rules. Structural fire design; permission to reproduce and derive extracts from British and ISO standards is granted by BSI. British Standards can be obtained in PDF or hard copy formats from the BSI online shop: www.bsigroup.com/Shop or by contacting BSI Customer Services for hardcopies only: Tel: +44 (0)20 8996 9001, Email: cservices@bsigroup.com.)

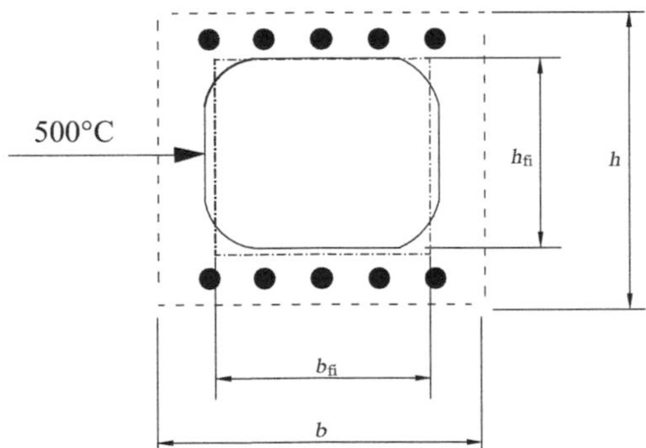

Figure 4.6 Reduced cross section of reinforced concrete beam and column with four-side fire exposure. (Reproduced from Figure B.1 of Page 72 BS EN 1992-1-2:2004+A1 (2019), Eurocode 2. Design of concrete structures, Parts 1–2: General rules. Structural fire design; permission to reproduce and derive extracts from British and ISO standards is granted by BSI. British Standards can be obtained in PDF or hard copy formats from the BSI online shop: www.bsigroup.com/Shop or by contacting BSI Customer Services for hardcopies only: Tel: +44 (0)20 8996 9001, Email: cservices@bsigroup.com.)

4.4.4 Concrete cover and protective layers

Concrete cover can enhance the fire resistance of concrete members. Eurocode has a detailed requirement of minimum cover thickness for concrete members in fire conditions.

In addition, as mentioned earlier, extra protective layers can also be used as a thermal barrier. The properties and performance of the materials used for protective layers should be assessed by appropriate test procedures.

4.5 DESIGN OF STEEL MEMBERS IN FIRE

4.5.1 Thermal response of steel in fire

As shown in Chapter 3, the thermal and mechanical properties of steel change in elevated temperatures, and the load-bearing capacity of steel decreases dramatically. To be able to design a structural member in fire, Eurocode specifies a temperature domain design method, called critical temperature method. It will be introduced here.

4.5.2 The critical temperature method (BS5950, 2003 and EN 1993-1-2 2005)

The critical (limiting) temperature of a member in a given situation depends on the load that the member carries under fire condition. Along with the increase of the fire, the limiting temperature is dependent on the fraction of the ultimate load capacity that a member has at the time of fire. The temperature causing the failure of a structural member depends on the utilization of the member in the fire situation. This is the simplest method of determining the fire resistance of a loaded member in fire conditions. However, when the load ratio is greater than 1, the member will fail at ambient temperature; this is because the structural member is overloaded. So, it failed primarily due to mechanical load rather than fire.

4.5.2.1 Assumptions

Eurocode 4 specifies the following assumptions for using critical temperature method:

- The temperature of the steel section is assumed to be uniform.
- The method is applicable to symmetric sections with a maximum depth of 500 mm, and the slab thickness it supports is >120 mm.

4.5.2.2 *Load ratio (degree of utilization)*

In design codes such as British code BS 5950 Part 8 (BSI, 2003c) and European standards (EN 1991-1-2, 2004; EN 1993-1-2, 2005; EN 1994-1-2, 2005), the structural member at the fire limit state is assessed according to the level of load. It is called load ratio or degree of utilization. It is the resistance of applied load of the structural members at the time of fire compared to that at the ambient temperature. The concept of load ratio is the basis for the limiting temperature method for steel structures in BS 5950 Part 8 (BSI, 2003) and critical temperature method in Eurocode 3 (ENV 1993-1-2/4.2.4, 2005a).

In BS5950 Part 8 (BSI, 1990), load ratio is defined as applied load (primarily due to dead load and live load) in fire conditions to those used in the design of the member at room temperature. In Eurocode 3 (ENV 1993-1-2 2/4.2.4, 2005), a similar ratio, which is called as degree of utilization is defined using the following formula:

1. For member Classes 1–3 and for all tension members:

$$\mu_0 = \frac{E_{fi,d}}{R_{fi,d,0}} \tag{4.1}$$

   where
   $E_{fi,\,d}$ is the applied load under fire condition,
   $R_{fi,\,d,\,0}$ is the design moment of resistance of the member at ambient temperature.

2. Alternatively, for tension members and for beams where lateral-torsional buckling is not a potential failure mode, it is conservative to use:

$$\mu_0 = \eta_{fi}\left[\frac{\gamma_{M,fi}}{\gamma_{M0}}\right] \tag{4.2}$$

   where
   $\eta_{fi,\,d}$ is the reduction factor.

4.5.2.3 *Critical temperature method for constrained members*

The critical temperature is the temperature at which failure is expected to occur in a structural steel element with a uniform temperature distribution. In Eurocode 3 (BS EN 3-1-2/4.2.4, 2005), the critical temperature is determined from

$$\theta_{cr} = 39.19\ln\left[\frac{1}{0.9674\mu_0^{3.833}} - 1\right] + 482 \tag{4.3}$$

where

μ_0 is the degree of utilization as shown in Equation 4.1 (or called load ratio),

It should be noticed that the above formula is primarily used for non-slender sections (Classes 1–3). For slender sections (Class 4), it is conservative to use 350°C as the critical temperature (as shown in Figure 4.7).

According to Eurocode3 (EN 1993-1-2/4.2.4, 2005), this equation can be used only for member types for which deformation criteria or stability considerations do not have to be taken into account (such as beams). This allows its use for tension members and restrained beams but precludes its use for both columns and unrestrained beams, where stability phenomena must be considered.

4.5.2.4 Critical temperature method for the compression and unconstrained members

Eurocode 3 (EN 1993-1-2/4.2.4, 2005) also provides the way to work out the critical temperature for compression members (such as columns) and unconstrained members, which is listed in Table 4.2.

4.5.2.4.1 Load ratio for column

For columns in simple construction exposed up to four sides, the load ratio is as follows:

$$U_0 = \frac{F_f}{A_g P_c} + \frac{M_{fx}}{M_b} + \frac{M_{fy}}{p_y + Z_y}$$

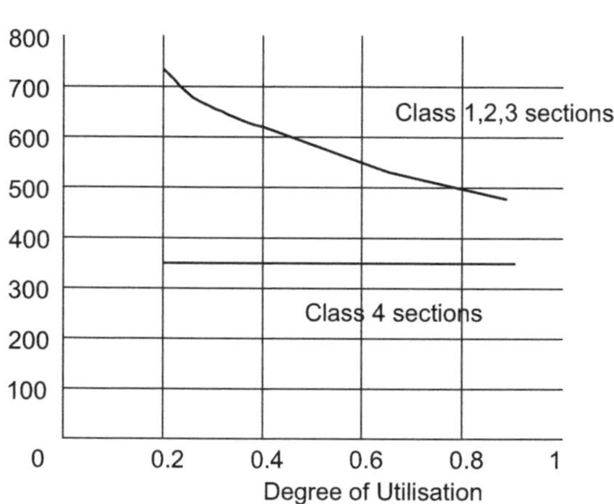

Figure 4.7 Relationship between critical temperature and degree of utilization for steel sections with different classes.

Table 4.2 Critical temperature of steel compression members

Compression member	*Critical temperature (°C) for utilization factor (load ratio)*					
$\bar{\lambda}_{\theta i}$	0.7	0.6	0.5	0.4	0.3	0.2
0.4	485	526	562	598	646	694
0.6	470	518	554	590	637	686
0.8	451	510	546	583	627	678
1	434	505	541	577	619	672
1.2	422	502	538	573	614	668
1.4	415	500	536	572	611	666
1.6	411	500	535	571	610	665

Partially adapted from Eurocode BS EN1993-1-2/NA.2.6 (2005a)

where
A_g is the gross area,
p_c is the compressive strength,
p_y is the design strength of the steel,
Z_y is the elastic modulus about the minor axis,
M_b is the lateral torsional buckling resistance moment,
F_f is the axial load at the fire limit state,
M_{fx} is the maximum moment about the major axis at the fire limit state,
M_{fy} is the maximum moment about the minor axis at the fire limit state.
For tension members exposed to four sides of fire,

$$U_0 = \frac{F_f}{A_g P_y} + \frac{M_{fx}}{M_{cx}} + \frac{M_{fy}}{M_{cy}}$$

where
M_{cx} is the moment capacity of the column about the major axis at the fire limit state,
M_{cy} is the moment capacity about the minor axis at the fire limit state.

4.5.2.4.2 Column slenderness in fire

In Eurocode 3, the column slenderness in fire can be expressed as follows:

$$\bar{\lambda}_{\theta i} = \bar{\lambda}\sqrt{\frac{k_{y,\theta}}{k_{E,\theta}}}$$

where
$\bar{\lambda}$ is the column slenderness,

$$\bar{\lambda} = \lambda\sqrt{\frac{p_y}{\pi^2 E^2}} \text{ with } \lambda = \frac{L_e}{r_y}$$

$k_{y,\theta}$ is the reduction factor from Section 3 for the yield strength of steel at the steel temperature θ_a reached at time t,

$k_{E,\theta}$ is the reduction factor from Section 3 for the slope of the linear elastic range at the steel temperature θ_a reached at time t.

4.5.2.5 Column buckling resistance in fire

The design of buckling resistance of column in fire a Class 1, Class 2, or Class 3 steel sections with a uniform temperature can be determined by

$$N_{b,fi,Rd} = \frac{\mathcal{X}_{fi} A k_{y,\theta} f_y}{r_{M,fi}}$$

where

$\mathcal{X}_{fi}$ is the reduction factor for flexural buckling in the fire design situation,

$k_{y,\theta}$ is the reduction factor from Section 3 for the yield strength of steel at the steel temperature Θ_a reached at time t,

$\mathcal{X}_{fi}$ is determined

$$\mathcal{X}_{fi} = \frac{1}{\varphi_\theta + \sqrt{\varphi_\theta^2 - \bar{\lambda}_\theta^2}}$$

with $\varphi_\theta = \frac{1}{2}\left[1 + \alpha\bar{\lambda}_\theta + \bar{\lambda}_\theta^2\right]$

and $\alpha = 0.65\sqrt{\frac{235}{f_y}}$

The axial compression force for column in fire can be calculated as follows:

$$P_{c,fi} = \frac{\mathcal{X}_{fi}}{1.2} P_{u,fi}$$

And

$$P_{u,fi} = A_s k_{y,\theta} p_y$$

4.5.3 Lateral torsional buckling of steel beams

In certain circumstances, lateral torsional buckling needs to be considered for steel beams under fire condition. In Eurocode 3, lateral torsional buckling is calculated by the same equation at ambient temperature with inclusion of temperature effect.

The beam slenderness in fire is given by

$$\bar{\lambda}_{LT,\theta} = \bar{\lambda}_{LT}\sqrt{\frac{k_{y,\theta}}{k_{E,\theta}}}$$

where

$\bar{\lambda}_{LT}$ is the beam slenderness at ambient temperature.

So, the lateral torsional buckling moment of beam under fire is

$$M_{b,\theta} = \frac{\chi_{LT,\theta}}{1.2} M_{p,\theta}$$

where

$M_{p,\theta}$ is the plastic bending resistance of the beam in fire.

And

$$\chi_{LT,\theta} = \frac{1}{\varnothing_{LT,\theta} + \sqrt{\varnothing_{LT,\theta}^2 - \bar{\lambda}_{LT,\theta}^2}}$$

$$\varnothing_{LT,\theta} = 0.5\left[1 + \alpha_{LT}\left(\bar{\lambda}_{LT,\theta} - 0.2\right) + \bar{\lambda}_{LT,\theta}^2\right]$$

4.5.4 Beams in line with compartment walls

When a beam is in line with a compartment wall, either above or supporting the wall, fire protection is required to limit deformation of the wall for the sake of integrity of the compartment (see Chapter 5).

4.6 MOMENT CAPACITY APPROACH (SECTION METHOD)

The superseded BS 5950: Part 8 (2003) specifies a calculation method for finding the fire resistance of members in bending. It is called "moment capacity method" or "Section Method" in some literatures. Though the code was superseded, the proposed method has been widely used in some references (Cai, 2020b; Han et al., 2007; Xiang et al., 2010; Yang et al., 2002). This method is easy for design engineers to use. Therefore, it will be introduced in this section.

4.6.1 Method of calculation

4.6.1.1 Temperature profile

The temperature distribution of the section at appropriate fire resistance periods must be known. Temperatures may be determined by fire tests or finite element modeling (as shown in Figure 4.8). Nowadays, it is primarily through heat transfer analysis available in most of the finite element analysis software.

4.6.1.2 Reduced strength of each element

After temperature profile of the section is determined, the member is divided into a set of elements of approximately equal temperature (as shown in Figure 4.8). The number of elements will depend on the complexity of the member and its temperature gradient. The reduced strengths of the various elements of the cross section can be calculated at elevated temperatures. The strength reduction can be worked out using strength deduction factors from Eurocode, as it has been introduced in Chapter 3.

4.6.1.3 Reduced flexural strength calculation

First, the "plastic" neutral axis of the section should be determined based on the reduced strength of each element. The moment capacity of the section then follows by multiplying the reduced strength of each element by the distance from the neutral axis and summing all the elements in the section.

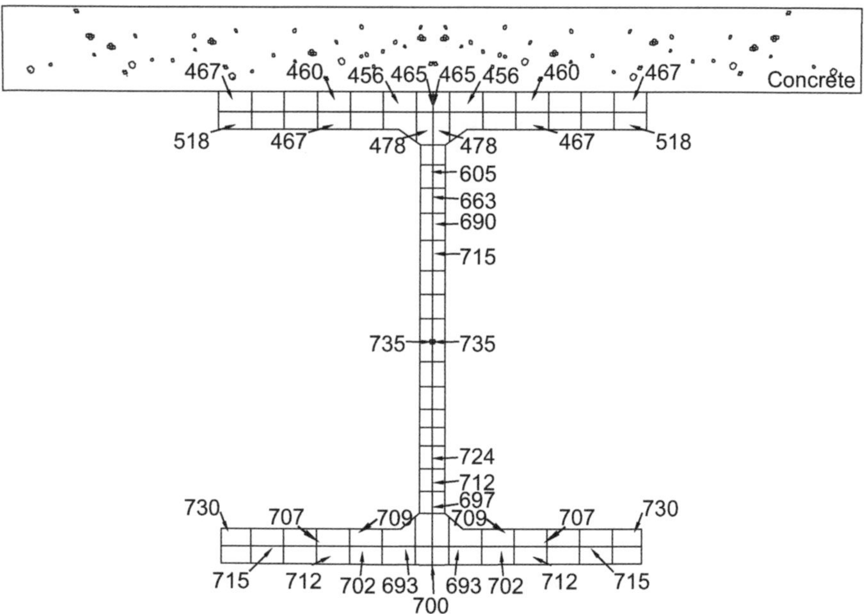

Figure 4.8 Element division and temperature profile of a typical composite beam section.

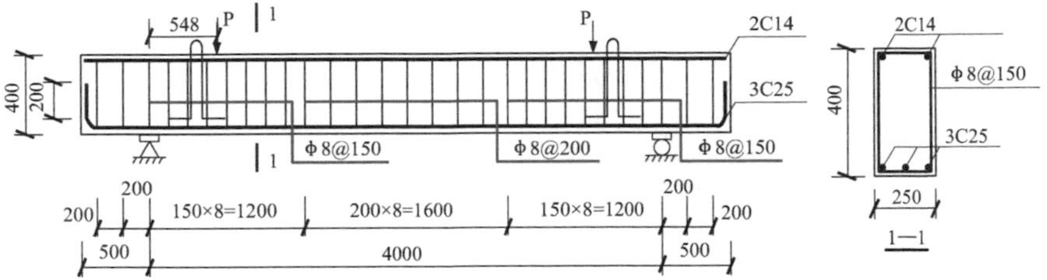

Figure 4.9 The dimensions and reinforcement of a reinforced concrete beam.

4.6.2 Case study for flexural capacity of reinforced concrete beams using moment capacity approach

In this section, based on the research of the author (Cai, Fu, et al., 2019), a case study of flexural capacity of reinforced concrete beams using moment capacity approach is shown here. As shown in Figure 4.9, a simply supported RC beam is exposed to fire. To obtain temperature profile of the RC beams, heat transfer analysis can be performed using ABAQUS. The specific heat and thermal conductivity, convection, radiation, and density of the materials should be first determined according to EN 1993-1-2 (2005), Eurocode 3. As shown in Figure 4.10, the temperature of concrete is obtained from the Abaqus model.

The beam section is then divided as shown in Figure 4.11 with temperature T_i in the ith cell.

The compressive strength reduction factor of concrete is obtained by substituting T_i into compression reduction factor of concrete in Equation 4.4. The temperature of the steel bar takes the highest temperature of the cell where the steel is located. The yield strength reduction factor from EN 1993-1-2 (2005), Eurocode 3 of reinforcement for the yield strength of the steel bar in high temperature is adopted.

$$\overline{\Psi}_{cT} = \frac{\sum \Psi_{cTi} \Delta b \Delta c}{b x_c} \tag{4.4}$$

where

b is the width of the beam section,
x_c is the height of the compression zone of the concrete beam section.

The post-fire residual flexural capacity is determined using the moment capacity method. The height of the compressive zone (x_c) of the concrete beam section after the fire can be calculated based on Equation 4.5:

$$f_c \sum \Psi_{cTi} \Delta b \Delta x + \Psi'_{yT} f'_y A'_s = \Psi_{yT} f_y A_s \tag{4.5}$$

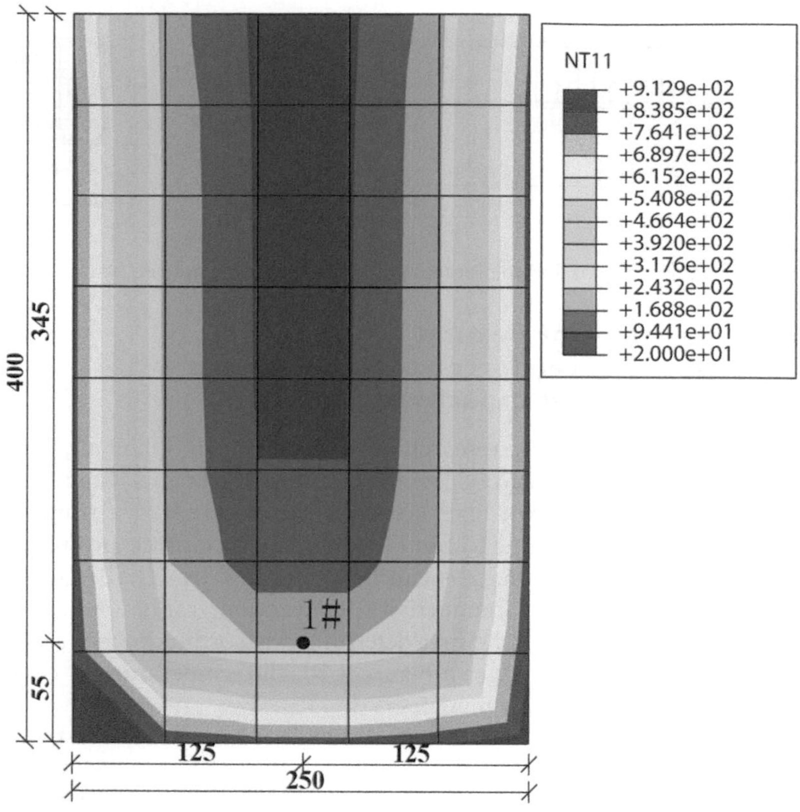

Figure 4.10 ABAQUS simulation result of temperature profile of concrete section after 1 h of fire exposure.

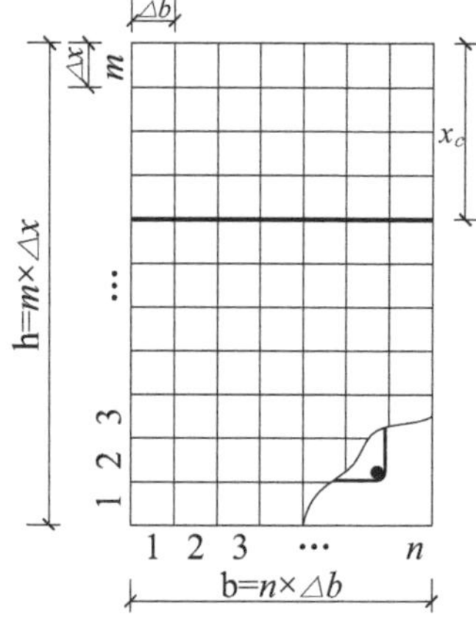

Figure 4.11 Section unit division of beam.

where

Ψ_{cTi} is the compressive strength reduction factor of the ith cell,

Δb is the width of a single cell,

Δx is the height of a single cell,

Ψ'_{yT} is the yield strength reduction factor of compressive steel,

f'_y is the yield strength of compressive steel at room temperature, N/mm^2,

A'_s and A_s are the area of steel in the compressive and tensile zone, respectively, mm^2.

After obtaining the height of the compression zone (x_c), the residual flexural capacity (M_u) of the RC beams after fire can be obtained as follows:

$$M_u = \alpha_1 \overline{\Psi}_{cT} f_c b x_c \left(h_0 - \frac{x_c}{2} \right) + \Psi'_{yT} f'_y A'_s \left(h_0 - a'_s \right)$$

where

α_1 is the coefficient of equivalent rectangular stress figure in compression zone of concrete, which is 1 here;

h_0 is the effective height of the beam section;

a'_s is the distance from the resultant force point of the compressive steel reinforcement to the margins of the compressive section.

4.6.3 Flexural capacity of steel beams using moment capacity approach

For designing steel member in fire, the code restricts the use of the moment capacity method to sections which are "compact" as defined in *BS 5950: Part 1*. It is assumed that the member fails in a flexural manner, without the occurrence of premature shear or instability effects in fire. In fire, the position of the "plastic" neutral axis changes when fire temperature changes. The lower parts of the cross section are normally exposed to the fire; therefore, the strength of lower parts is weakened, and hence the plastic neutral axis shifts upwards. For simple beams supporting concrete floors, the plastic neutral axis can rise close to the upper flange (as shown in Figure 4.12).

In the moment capacity method, the strength reduction factor is applied to the original strength of the steel at different locations across the section on the basis of temperature distribution by finite element analysis. The section modulus of the beam in fire is therefore determined by considering the moment contribution of the stress blocks. It is normally found that the ratio of the section modulus of an I section in fire to that in normal design increases to a value between 1.2 and 1.5 depending on the proportions of the cross section.

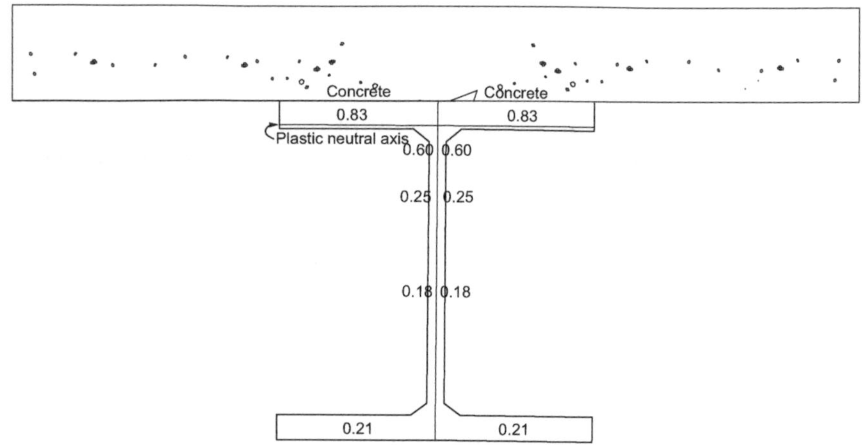

Figure 4.12 Position of plastic neutral axis and deducted stress distribution after fire.

4.7 DESIGN OF COMPOSITE BEAMS UNDER FIRE

Composite beams are widely used in the steel-framed tall buildings. Figure 4.13 shows a typical steel composite building under construction.

Composite floor systems have many forms such as composite beams with metal decking slabs (a most widely used in the construction). Figure 4.14 shows another type of composite beam which precasts hollow-core slabs as floor system.

4.7.1 Resistance of shear connection in fire

Composite beams utilize the composite action between concrete slabs (in most cases metal decking slabs) and steel beams through shear connectors (as shown in Figure 4.15). Therefore, the behavior of shear connectors in fire will also affect the behavior of the composite action, and therefore the flexural capacity of composite beams in fire.

The design formulas for shear resistance of shear studs in fire are given in EN 1994-1-2 (2005) Eurocode 4:

$$P_{fi,\,Rd} = 0.8k_{u,\,\theta}P_{Rd}$$

with P_{Rd} as obtained from Equation 6.18 of EN 1994-1-1.

$$P_{fi,\,Rd} = k_{c,\,\theta}P_{Rd}$$

with P_{Rd} as obtained from Equation 6.18 of EN 1994-1-1.

Figure 4.13 A typical composite buildings. (Photo taken by the author.)

Figure 4.14 Composite beam with precast hollow-core slabs. (Photo taken by the author.)

Figure 4.15 Shear connectors in composite beams.

where
values of $k_{u,\theta}$ and $k_{c,\theta}$ are taken from Tables 3.2 and 3.3 of EN 1994-1-1, respectively.

The temperature θ_v ^{0}C of the stud connectors and θ_c ^{0}C of the concrete material may be taken as 80% and 40%, respectively, of the temperature of the upper flange of the beam.

4.7.2 Effect of degree of shear connection

As it is widely known, the degree of shear connection greatly affects the capacity of the composite beams. BS 5950-8 (2003) specifies the effect of the different degree of shear interactions. Steel Construction Institute (SCI) P288 (2000) is represented in Table 4.3.

Table 4.3 Limiting (Critical) temperature for composite beams with different degree of shear interactions

Degree of shear connection (%)	*Limiting temperature (°C) for a load ratio of*						
	0.9	*0.8*	*0.7*	*0.6*	*0.5*	*0.4*	*0.3*
100	485	520	550	580	610	645	685
40	525	550	575	600	635	665	700

SCI Publication P288 (2000), *Fire Safety Design: A New Approach to Multi-Storey Steel Framed Buildings.*

4.7.3 Edge beams in fire

As it is well known, the edge beam is designed as non-composite in most of the projects. But in fire design, the floor slab should be adequately anchored to the edge beam to ensure membrane action in the slab. Therefore, if it is designed to be non-composite, it must still provide shear connectors at minimum 300 mm spacing with steel mesh of the slab to be hooked to the shear connectors (SCI Publication P288, 2000).

In addition, edge beams are supports for cladding, and to avoid fall of debris (such as broken claddings), the deformation during the fire should be controlled. Therefore, all the edge beams need to be fire protected. In addition, vertical ties and wind post can be used to further restrain the deformation.

4.7.4 Case study of composite beam design in fire

The stress and strain distribution of a composite beam in fire conditions is illustrated in Figure 4.16. At the critical temperature, the lower flange of the beam is fully yielded. all of the steel section is in tension, making the plastic neutral axis of the composite section usually lie in the concrete slab. Therefore, in fire condition, higher bottom flange strains are generated. Therefore, a strain limit of 2% is recommended by SCI (1990) when assessing the moment capacity of the section, subject to the "stickability" of the fire protection if the member is fire protected.

4.8 DESIGN OF COMPOSITE SLABS IN FIRE

Composite slab's floor system has been widely used in tall buildings due to its good span to depth ratio. Figure 4.17 shows a typical metal decking composite floor spanning over a secondary beam. It is comprised of cast in situ concrete, steel mesh, shear connector, metal decking, and beams.

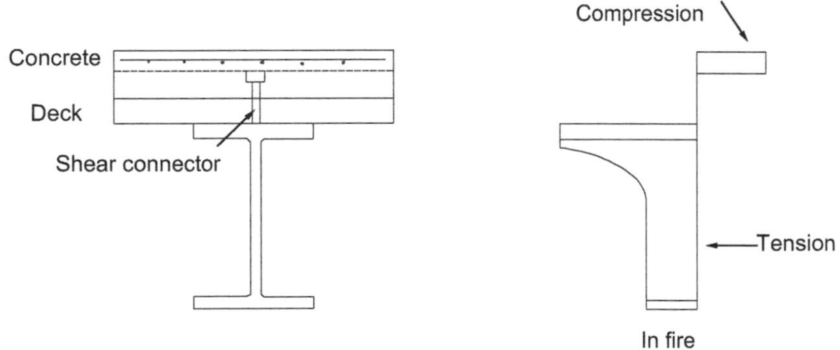

Figure 4.16 Stress distribution of composite beams in fire (SCI, 1990).

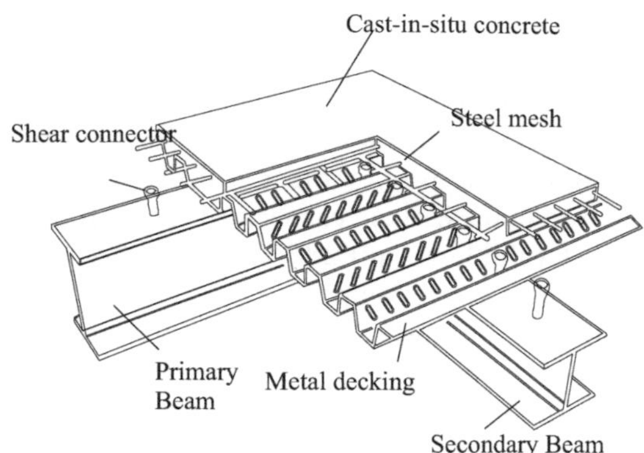

Figure 4.17 A typical composite floor system.

All these components play important roles during the fire. The role of the shear connectors has been introduced in the last section. Metal decking also improves the integrity of the whole slab; in the meantime, it provides a shielding effect to the slab, shielding the heat flow into the concrete, and controls spalling.

4.8.1 Membrane actions in fire

The results from Cardington tests show that composite slab plays a crucial role in fire. The behavior of composite slab in fire shows significant difference to its normal condition. In room temperature, they span in one direction along the metal decking and behave as a one-way slab. However, in fire, the slab acts as a membrane supported by the cooler perimeter beams and protected columns. Tensile membrane action is a load-bearing mechanism when slabs undergoing large vertical displacement, where the induced radial tension in the center of the slab is sustained by a peripheral ring of compression. A diagrammatic representation of tensile membrane action is provided in Figure 4.18.

If the perimeter of the slab is fixed supported, compression membrane action starts first at small deflection stage. As the deformation becomes greater, tensile membrane action starts and loads are primarily carried by the reinforcement mesh. To enhance the tensile membrane action, the steel mesh in the slabs needs good anchorage to the supports.

4.8.2 Strength design composite slabs

The moment capacity method can be used for the design of composite slabs under fire. The thermal profile shall be worked out, from which the reduced strength of the concrete and steel can be calculated. As a simplification of the actual behavior, the strength reduction factors may be taken according

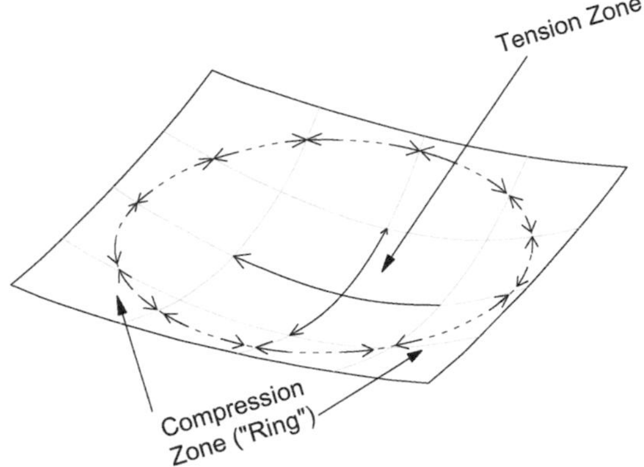

Figure 4.18 Tensile membrane action in slab.

to the Eurocode. The moment capacities of the section in hogging, sagging, and bending can then be evaluated.

Due to the complexity of working out the temperature and reduced strength of the metal decking, it is conservative to ignore the tensile capacity of the deck. Therefore, the calculation can be greatly simplified.

For continuous floors, their capacities can be combined by considering the plastic failure mechanism of the floor. It is noticeable that there is a redistribution of moment during a fire from the mid-span area to the supports.

4.8.2.1 Calculation method based on plastic theory

SCI Publication 056 (1991) introduces formulas for plastic failure of a continuous composite metal decking slabs in fire.

For middle span

$$M_h + M_s \geq M_0$$

where

M_h=hogging moment capacity in fire per unit width,
M_s=sagging moment of capacity in fire per unit width,
M_0=free bending moment per unit width.

$$= \frac{L^2}{8}\left(\gamma_{fd} w_d + \gamma_{fi} w_i\right)$$

For end span

$$M_p + 0.5M_n\left(1 - \frac{M_n}{8M_0}\right) \geq M_0$$

where

M_p=sagging,

M_n=hogging moment capacity of the composite section,

M_0=free bending moment applied to the simply supported slab in fire conditions.

The second term approximates to a value of 0.45 M_n. For an internal span of a continuous slab, the plastic moment capacity is simply

$$M_n + M_p \geq M_0$$

4.8.2.2 Calculation method considering membrane action

Bailey et al. (2000a, b) and Bailey (2001) developed a calculation method for determining the ultimate load-carrying capacity of two-way slabs incorporating the effects of tensile membrane enhancement under elevated temperatures. The design method calculates an enhancement factor due to effects of the membrane forces on the flexural strength. The method considers the failure mode shown in Figure 4.19. This method has been adopted by the SCI.

In designing the slab, Bailey (2000) divides a composite floor into several horizontally unrestrained rectangular fire-resisting zones. These are composed internally of simply supported unprotected beams (Bailey and Moore, 2000a). With increasing exposure to elevated temperatures, the formation of plastic hinges in the unprotected beams redistributes the loads to the two-way bending slab, undergoing large vertical deflection. The design method also considers ultimate failure of the system based on the maximum permissible deflections due to the mechanical strains of the reinforcement and the thermal bowing deflections.

Based on the diagram in Figure 4.20, the ultimate vertical deflection at the fire limit state is derived from a combination of thermal bowing of the slab and the mechanical strain in the reinforcement, which is defined in the following:

$$v = \frac{\alpha(T_2 - T_1)}{19.2h} + \sqrt{\left(\frac{0.5f_{y,\theta}}{E_{,\theta}}\right) \times \frac{3L^2}{8}} \qquad (4.16)$$

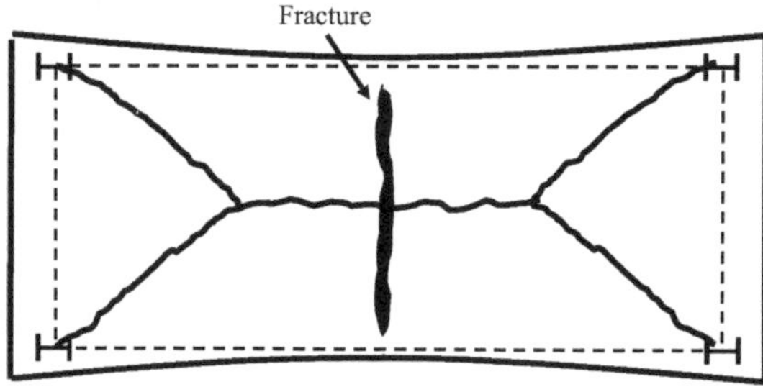

Figure 4.19 Failure mode of slab in fire due to membrane action.

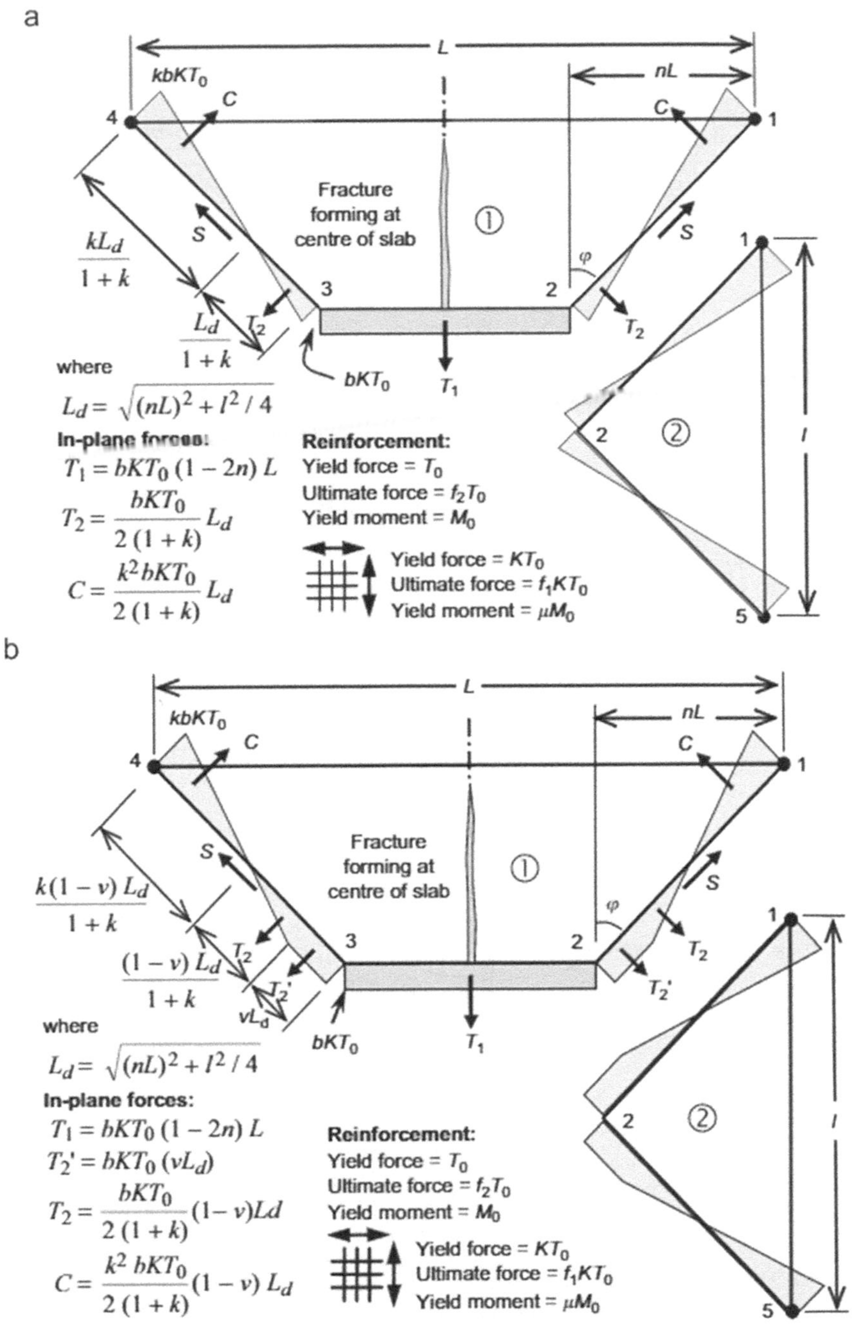

Figure 4.20 In-plane stress distribution patterns by Bailey (2007).

The deflection due to mechanical strain of the reinforcement is limited to span/30 (see Section 4.9.4 for further details).

The composite slab capacity at any given time in fire is calculated as

$$w_{p\theta} = e\left(\frac{\textit{Internal work done by the composite slab in bending}}{\textit{External work done by the applied load per unit load}}\right)$$

$$+\left(\frac{\textit{Internal work done by the beam in bending}}{\textit{External work done by the applied load per unit load}}\right)$$

where

$w_{p\theta}$ is the slab panel capacity at a given time,

e is the enhancement of the slab capacity, calculated as in the reference (Bailey and Toh, 2007).

4.8.3 Insulation criterion of composite slabs

BS 476: Part 20 (2014) specifies that the composite slabs should have the ability to limit the conduction of the heat to the upper surface. The average temperature raise of the upper surface should not exceed 140°C or a peak temperature of 180°C as shown in Figure 4.21. In determining temperatures

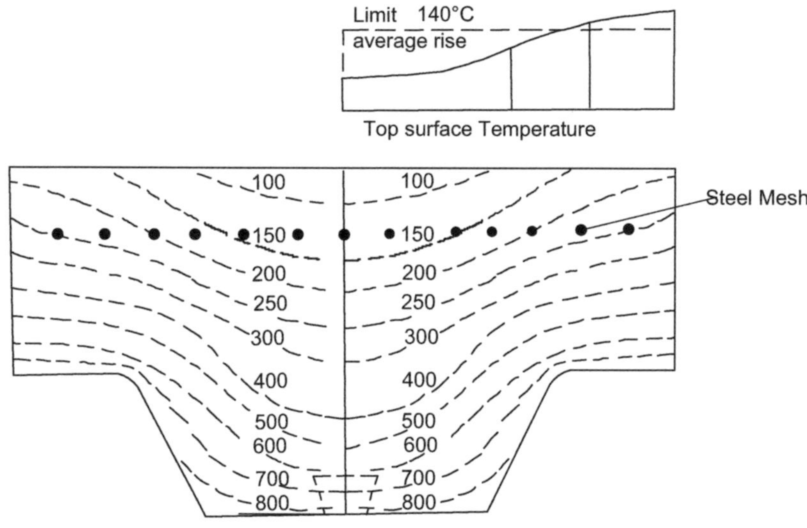

Figure 4.21 A typical temperature profile of composite metal deck slab with the average temperature rise of the top surface <140°C.

of the upper surface of the slab, account should be taken of the "dwell" or delay in temperature at around 100°C resulting from the vaporization of free moisture.

This insulation criterion is achieved through the requirements of minimum insulation thickness and minimum slab depths. The minimum slab depth is considered to be the depth of the concrete topping. For composite deck slabs comprising trapezoidal profiles, the "heat-sink" effect of the closely spaced ribs means that minimum slab depths can be relaxed.

4.8.4 Deformation design of composite slabs in fire

The deformation of the floor shall not cause the failure of the compartmentation. Therefore, the limitation of deformation of composite slabs is specified for the sake of integrity of the compartment (see Chapter 5). BS 476: Part 20 (2014) requires that the deflection of the composite slab be limited to span/20 and the rate of the deflection R (mm/minute).

$$R < \text{Span}^2/9{,}000d \tag{4.17}$$

where

d is the distance from bottom of the tension zone to the top surface of the slab.

4.9 DESIGN OF POST-TENSION SLABS IN FIRE

As shown in Figure 4.22, post-tension slabs have been widely used in most of the tall buildings, such as the Shard. This is because post-tensioned floor slabs can result in thinner concrete sections and longer spans between supports, and thus reduce the storey height, which is important for tall buildings, as reducing overall height of the building can dramatically reduce the cost of the project (Fu, 2018).

Post-tensioned concrete slabs can be constructed using unbonded or bonded tendons. For unbonded slabs, the transfer of force from the tendons to the concrete is via the end anchor. In bonded slabs, the transfer of the force is via the end anchors, together with the bond between concrete and the tendons. So far, there is no clear guidance on how to design post-tensioned slabs in fire. Bailey and Ellobody (2009) performed eight fire tests on bonded post-tensioned one-way concrete slabs. In all these tests, cracks occurred directly in line and parallel to the tendons due to thermal stresses at relatively low tendon temperatures, which were not

Figure 4.22 Post-tensioning slabs in construction. (This file is licensed under the Creative Commons Attribution-Share Alike 3.0 Unported license. Permission is granted to copy, distribute, and/or modify this document under the terms of the GNU Free Documentation License, Version 1.2 or any later version published by the Free Software Foundation; with no Invariant Sections, no Front-Cover Texts, and no Back-Cover Textshttps://commons.wikimedia.org/wiki/File:Post-Tensioning-Cables-2.jpg.)

observed in the ambient tests. It was shown that the use of plastic ducts resulted in slightly higher tendon temperatures due to the ease at which water migrated from the grout once the duct had melted. The fire tests have shown that the fire resistance specified in current codes of practice is generally conservative for bonded post-tensioned one-way spanning concrete slabs.

Gustaferro (1973) made a report on an analysis of 18 full-scale fire tests of concrete slabs and beams prestressed by post-tensioning. He recommended that

- The cover for post-tensioned tendons in slabs should be the same as the cover for reinforcing steel in slabs. These criteria can be applies to slabs with post-tensioned tendons.
- The cover to the prestressing steel at the anchor should be at least ¼ in. greater than that required away from the anchor. Minimum cover to the steel bearing plate should be at least 1 in. in beams and 3/4 in. in slabs.

4.10 DESIGN OF CONNECTIONS UNDER FIRE

The results of Cardington tests show that in fire, the connections have the potential to develop additional moments and, therefore, enhance the load-bearing capacity of the beam. It is found from these tests that the connections perform better in heating phase than in cooling phase. This is because the plastic compression deformation of the beam due to the restrain during the expansion causes permanent shortening of the beam in the cooling stage. The connections subjected to tension force in cooling may lead to failure of the connection. Therefore, in selecting the type of the connection, the ability of the connection to deform well is an important factor that needs to be considered.

New Zealand fire design code NZS 3404 Parts 1 and 2 (1997) require that the connections linked to protected members have fire protection with the same thickness as the maximum thickness of the members framing into the connection. This thickness should be maintained over the entire section of the connection including bolt heads, welds, and splice plates. This is a conservative fire safety design approach for connections.

Connections that transfer design actions from a member require a fire resistance rating. The fire resistance of connection components can be calculated using the critical temperature method introduced in this chapter. Connectors shall achieve the same value as that for the member being supported, ensuring that failure of the connection does not occur.

The Eurocode does not give guidelines for the fire resistance rating of connections, but since it gives details on the design of the members, this theory can be transferred to the design of connections.

4.11 DESIGN OF BEAM OPENINGS

The Eurocode 3 Document and the British Steel code do not have recommendations on the fire resistance of beam openings. NZS 3404 states that unless determined in accordance with a rational fire engineering design, the thickness of fire protection material at and adjacent to web penetrations shall be the greatest of

- That required for the area of beam above the penetration considered as a three-side fire exposure condition;
- That required for the area of beam below the penetration considered as a four-sided fire exposure condition;
- That required for the section as a whole considered as a three-side fire exposure condition. It also specifies that the thickness shall be applied over the full beam depth and shall extend each side of the penetration for a distance at least equal to the beam depth and not less than 300 mm.

4.12 SUMMARY OF STRUCTURAL FIRE DESIGN METHODS

4.12.1 Comparison of moment capacity method and critical temperature method

The moment capacity method will usually be significantly lower than that obtained by the critical temperature method. This is because, in critical temperature approach, when simple beams supporting concrete slabs are subjected to fire, the temperature distribution of a steel I section is considered to be uniform. However, in reality, the lower portion of the web and lower flange are fully exposed. The upper-flange temperature of an unprotected beam is 150°C–200°C lower than that of the lower flange at its critical temperature. Therefore, the moment capacity method has a lower capacity than that of critical temperature method.

Other reasons for this conservatism are as follows:

- On average, steel will be at least 10% stronger than the characteristic value used in normal design and that steel strains considerably higher than 1.5% will be experienced in tests. Therefore, the moment capacity method results in a low value.
- Temperatures are often lower than those obtained in finite element modeling.
- The interaction between the elements may offer more restraint or strength.

In principle, the moment capacity method may be applied to any structural form, but it has proved to be the most appropriate for determining the capacity of partially protected beams and floors in fire.

4.12.2 Comparison of three major methods

In this chapter, several structural fire design methods have been introduced. The simplest design method is a critical temperature method by Eurocode. Moment capacity method is more complicated. The most accurate method is using numerical software to design structural member in fire (will be introduced in Chapter 6). As shown in Figure 4.23, numerical analysis is the most accurate way in structural fire design. However, the degree of knowledge to an engineer and degree of complexity increase. Depending on the complexity of the project and the availability of the numerical software, an engineer can make a choice between these three methods.

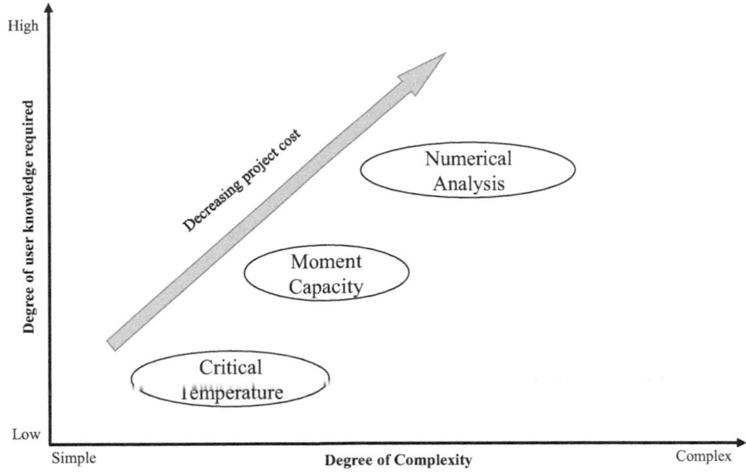

Figure 4.23 Comparison of three structural fire design methods.

REFERENCES

Bailey, C. G., Ellobody, E. (2009), Fire tests on bonded post-tensioned concrete slabs. *Engineering Structures*, 31(3), pp. 686–696

Bailey, C. G., Moore, D. B. (2000a), The structural behaviour of steel frames with composite floor slabs subject to fire: Part 1: Theory. *The Structural Engineer*, 78(11), pp. 19–27.

Bailey, C. G., Moore, D. B. (2000b), The structural behaviour of steel frames with composite floor slabs subject to fire: Part 2: Design. *The Structural Engineer*, 78(11), pp. 28–33.

Bailey C. G., Toh W. S. (2007), Behaviour of concrete floor slabs at ambient and elevated temperatures. *Fire Safety Journal*, 42(2007), pp. 425–436.

BS 476-20:1987 (2014, April), Fire tests on building materials and structures. Method for determination of the fire resistance of elements of construction (general principles), AMD 6487 Corrigendum, The British Standards Institute.

BS 5950-8 (2003), (Superseded, Withdrawn) Structural use of steelwork in building. Code of practice for fire resistant design. The British Standards Institute.

BS EN 1992-1-2:2004+A1 (2019), Incorporating corrigendum July 2008, Eurocode 2. Design of concrete structures, Part 1–2: General rules. Structural fire design.

BSI 'PD7974-3 (2019), Application of fire safety engineering principles to the design of buildings, Part 3: Structural response and fire spread beyond the enclosure of origin (Sub-system 3)'. The British Standards Institute.

Cai, B., Wu, A., Fu, F. (2020a), Bond behavior of reinforced porous light-weight cinder concrete using PP fiber after fire exposure. *Concrete and Computer. An International Journal*, **26**(2), 115–125.

Cai, B., Li, B., Fu, F. (2020b). Finite element analysis and calculation method of residual flexural capacity of post-fire RC beams. *International Journal of Concrete Structures and Materials*, **14**(1).

EN 1993-1-2 (2005), Eurocode 3. Design of steel structures, Part 1–2: General rules. Structural fire design. Commission of the European communities.

EN 1994-1-2 (2005), Eurocode 4. Design of composite steel and concrete structures, Part 1–2: General rules. Structural fire design. Commission of the European communities.

Fu, F. (2018), Design and Analysis of Tall and Complex Structures. Elsevier. ISBN 978-0-08-101018-1.

Gustaferro, A. H. (1973), *Fire Resistance Ofpost-Tensioned Structures*, The Consulting Engineers Group, Inc., Glenview, IL.

Han, Y. L., Wang, Z. Q., Wang, Y. J., Bai, L. L. (2007), Analysis of bending capability of a reinforced concrete beam supported at both ends in fired field. *Journal of Naval University of Engineering*, 19(1), pp. 76–80.

HM Government (2019), The Building Regulations 2010-Approved Document B, Volume 1 fire safety, Dwellings', HM Government.

HM Government (2019), The Building Regulations 2010-Approved Document B, Volume 2 fire safety, Buildings other than dwellings', HM Government.

Khoury, G. A., Anderberg, Y. (2000), Fire safety design (FSD): Concrete spalling review. Report submitted to the Swedish National Road Administration, Sweden, 2000, pp. 37–40.

NZS 3404 Parts 1 and 2:1997, Steel Structures Standard. Standards Council, New Zealand.

OFFSHORE TECHNOLOGY REPORT2001/074 (2001), HSE Health & Safety Executive, Deterioration and spalling of high strength concrete under fire Prepared by Sullivan & Associates for the Health & Safety Executive

SCI 1990, Fire Resistant Design of Steel Structures-A Handbook to BS 5950: *Part* 8. The Steel Construction Institute Silwood Park, Ascot, Berkshire.

SCI Publication 056 (1991), *The Fire Resistance of Composite Floors with Steel Decking* (2nd Edition). The Steel Construction Institute Silwood Park, Ascot, Berkshire.

SCI Publication P288 (2000), *Fire Safety Design: A New Approach to Multi-Storey Steel Framed Buildings*. The Steel Construction Institute Silwood Park, Ascot, Berkshire.

Xiang, K., Wang, G. H., Yu, J. T., Wang, S., Diao, X. L. (2010), Uniform calculating method of flexural capacity for fire-damaged reinforced concrete bending members. *Fire Science and Technology*, 29(12), pp. 1035–1039.

Yang, J. P., Shi, X. D., Guo, Z. H. (2002), Simplified calculation of ultimate load flexural capcity of reinforced concrete beams under high temperature. *Industrial Construction*, 32(3), pp. 26–28.

Zhukov, V. V. (1975), *Explosive Failure of Concrete during a Fire (in Russian)*. Translation No. DT 2124. Joint Fire Research Organization, Borehamwood.

Chapter 5

Typical fire safety design strategy for tall buildings

5.1 INTRODUCTION

In this chapter, detailed design strategies for tall buildings is introduced. The prescriptive and performance-based fire design approaches are first introduced, followed by a fire risk analysis. The deterministic and probabilistic approaches to determine the worst-case fire scenarios are then introduced. A detailed introduction of compartment design and evacuation route design for tall buildings is made followed by other design issues, such as firefighter access and fire protection requirement to façade. The fire alarm and communication, and fire and smoke suppression system are also discussed. At the end of this chapter, case studies of two real construction projects, Burj Khalifa and the Shard, are made.

5.2 FIRE SAFETY DESIGN OBJECTIVES AND STRATEGIES FOR TALL BUILDINGS

Fire safety design of tall buildings requires engineers to use relevant design measures to protect people and property from fire. When designing new buildings or renovating existing buildings, engineers develop the plan for fire protection. The following are the key goals of a fire safety design for tall buildings (SFPE, 2013):

1. Life safety
2. Property protection (such as stability of the building).

5.2.1 Design objectives

The following are the key design objectives of fire safety design for tall buildings:

- To contain a fire to its origin.
- To ensure that the occupants can be quickly evacuated from the building when fire happens.
- To design structures that withstand fire conditions, which include
    1. Resistance of building members to fire, mainly through fire protection measures
    2. Resistance to structural collapse, ensuring stability of the building
    3. Resistance to flame penetration, ensuring the integrity of the compartment.

5.2.2 Design strategies

To satisfy the aforementioned design goals and objectives, the main design strategies are as follows:

- **Prevention**: controlling fire ignition.
- **Warning**: ensuring that the occupants are informed in the event of fire.
- **Containment**: fire should be contained to the smallest possible area.
- **Evacuation**: ensuring that the occupants of buildings and surrounding areas are able to move to places of safety; ensuring also that adequate means of escape and protection of escape routes are provided.
- **Extinguishment**: ensuring that fire can be extinguished quickly and with minimal consequential damage.
- Sufficient structural fire design, such as fire protection regimes for key building members.

5.2.3 Design process

Fire safety design process primarily includes the following:

- Fire hazards potential analysis
- Fire scenarios analysis
- Compartment and evacuation route design
- Fire and smoke detection and suppression and communication systems
- Structural fire analysis.

These design processes will be introduced in this chapter in detail, except "structural fire analysis" which has been covered in Chapter 4.

5.3 DESIGN STRATEGY FOR TALL BUILDINGS IN FIRE

5.3.1 Prescriptive design

Prescriptive design codes have been used for fire safety design of tall buildings in the past. In prescriptive codes, most requirements prescribe the bound for the engineers. Therefore, it is easy to use in the design process by an engineer. However, for tall buildings, different factors such as plan layout, compartment sizes, and opening factors have great influence on fire scenarios. Different building characteristics result in different fire scenarios. Prescriptive design codes may lead to either under- or overdesign due to its inflexibility.

5.3.2 Performance-based design

Performance-based design uses a holistic approach, requiring interactions between all fire protection systems in the building. In performance codes, desired objectives are determined at the first place. The designers are given the freedom to choose a solution that will meet the objectives. Performance-based provisions can be found in recent codes and standards, including:

- International Fire Code (IFC)
- U.S. National Fire Protection Association (NFPA), Uniform Fire Code
- NFPA 101, Life Safety Code
- NFPA 5000, Building and Construction Code
- SFPE Engineering Guide to Performance-Based Fire Protection Analysis and Design of Buildings.

Performance-based design considers the whole building's behavior in fire while designing. It comprises several key procedures which will be introduced here.

5.3.2.1 Step 1: set fire safety goals and objectives

When using performance-based designs, fire safety goals for a building are fist identified (such as life safety and property protection) based on the importance, category, and occupancy of the building.

5.3.2.1 Step 2: determine performance criteria

The performance criteria of the building in fire can be divided into deterministic and probabilistic criteria.

Deterministic criteria include fire ignition criteria, fire growth criteria, fire resistance and spread criteria, and life safety criteria.

In the design, Under deterministic criteria, engineers require to quantify the fire processes such as fire growth, fire and smoke spread, evacuation time, structural response, and the consequences of these processes on the building and its occupants. The results of these calculations are then compared to established deterministic criteria to determine whether the design is satisfactory.

In terms of structural response, the fire resistance rating for each structural member needs to be determined. The performance of structural elements will be studied assuming that each of them is protected by different fire protection regimes. Acceptance of the suitability of structural system will be assessed based upon structural performance under different fire scenarios studied. In this method, A fire scenario analysis has to be made, as thermal responses of individual elements are affected by different atmosphere curve, and their structural response is influenced by individual thermal responses.

Probabilistic criteria are risk criteria established based on statistical data. Acceptable risk levels at a minimum cost will be set for the design. It estimates fire severity and likelihood of occurrence and then set up objectives in terms of reducing the likelihood of occurrence, the severity of the incident, or both.

In the design, engineers will use probabilistic risk assessment methods to make a probabilistic evaluation with a holistic view of the whole building (such as layout, occupants, category, and function of the building) and estimate risk levels of the likelihood of a fire incident occurring and its potential consequences like injury, death, etc. The calculated risk levels are then compared to the risk criteria to determine whether the proposed designs are acceptable. The probabilistic method will be introduced in detail in Chapter 6.

5.3.2.2 Step 3: analysis of fire scenarios

Analysis of fire scenarios determines the types of fires that are likely to occur. It is based on the characteristics of a building such as the materials used, layout, category, and function of the building. It is greatly affected by the design of the compartment and the opening factors.

5.3.2.3 Step 4: protection strategy

The types of fire protection strategies that are used in performance-based design are the same as those that are used in prescriptive codes, and it is comprised of detection, suppression, egress, and fire resistance. Quantitative assessment of design options against the fire safety objectives uses engineering tools, methodologies, and performance criteria (SFPE, 2013,).

5.3.2.4 Step 5: determine whether the fire safety goals are met

After fire protection strategies are developed, they are evaluated to determine whether the fire safety goals are met for each of the fire scenarios.

5.3.3 Summary

In most cases, the entire building need not to be designed based on its performance. The building can be partially designed using prescriptive codes. Performance-based design ensures that the fire performance of the whole building will be considered. Therefore, the fire protection-related systems must be coordinated with each other and with other building systems, which include the following (Hadjisophocleous and Benichou, 1999):

- Coordination of sprinkler system with fire alarm system zoning
- Coordination of sprinkler system water flow
- Coordination of fire alarm and egress system with building security
- Coordination of smoke control systems with detection and HVAC system designs
- Coordination of fire separations with architectural designs
- Coordination of penetrations of fire-rated assemblies with mechanical and electrical designs (e.g., piping, ductwork, and wiring penetrations)
- Coordination of means of egress with architectural designs.

5.4 FIRE RISK ANALYSIS FOR TALL BUILDINGS

Fire risk assessment is a function of the likelihood of fire and the consequences or severity.

$$\text{Risk} = \text{likelihood of fire} \times \text{consequence} \tag{5.1}$$

5.4.1 Qualitative fire risk assessment

Qualitative assessments are typically adopted when a simplistic issue or concern is identified. Furthermore, they have been utilized to rationalize or identify key elements within a fire scenario to provide an overview of the risk that may be present within a certain situation.

Through a more detailed qualitative assessment, the practitioner is able to assess those high-risk elements/scenarios, which would commonly be a quantitative assessment. A qualitative risk assessment includes the following steps: architecture review, building environment and occupant characterization, fire safety management, evacuation strategies, acceptance criteria, and fire hazards identification.

5.4.2 Quantitative fire risk assessment

A quantitative probabilistic risk assessment analysis includes hazard identification, consequence analysis, frequency analysis, uncertainties, and sensitivity analysis. Engineers first develop a number of fire scenarios, and then estimate the likelihood and consequences of these scenarios, followed by an evaluation of the risk based on Equation 5.1. At the end, a comparison is made between the calculated risks and the acceptable risks to find out whether the evaluation criteria are satisfied.

5.5 DETERMINISTIC AND PROBABILISTIC ASSESSMENTS TO DETERMINE THE WORST-CASE FIRE SCENARIO

As introduced in the preceding sections, for a tall building, the worst-case fire at the fire limit state needs to be determined. There are two potential methods of assessing a worst-case fire.

5.5.1 Deterministic approach

A deterministic approach uses the recommended values at each stage of the analysis as outlined in the various codes and frameworks. This method can achieve required structural reliability for a particular structure by the adoption of lower-bound values for the various inputs such as material properties and loadings. The worst-case fire can be determined by varying the key parameters of the fire to assess the sensitivity of response to the given inputs. The sensitivity of the key input variables of fire loading and ventilation conditions should be assessed for a sample of structural element types in a number of compartments.

5.5.2 Probabilistic approach

A probabilistic approach is based upon the statistical data of the key input values and an assessment of the likelihood of exceedance of a given limit state. It relies on the engineering model to allow many iterations of the model to deploy the likelihood of the occurrence of a worst-case fire for certain type of buildings. The probability of exceedance of a given limit state needs to be coupled with an assessment of the consequences of failure of that individual element of structure. A Monte Carlo simulation technique can be used in this approach. For example, the fire scenarios are primarily affected by two factors: opening factors and fire load density. Knowing the existing probabilistic distribution and the statistical parameters, the opening factor and fire load density can be sampled using Monte Carlo simulation; therefore, different fire scenarios can be generated. Refer to Chapter 6 for further details.

5.6 COMPARTMENT DESIGN

In a fire safety design, it is essential to ensure that the fire will be inhibited in its origin as much as possible. This is primarily achieved through the design of fire compartments. Approved Document B (2019) defines a fire compartment as: "A building or part of a building, comprising one or more rooms, spaces or storeys, that is constructed to prevent the spread of fire to or from another part of the same building or an adjoining building."

The design of fire compartment restricts the spread of fire by dividing a building into a number of compartments. These fire compartments are separated by fire-resistant building members, such as compartment walls and compartment floors. Internally, the compartment floors and walls serve as barriers for a designed fire resistance period as long as they retain their integrity. Externally, fire should not spread through the façade.

Figure 5.1 shows a typical compartment design for a tall building. It can be seen that the plan layout is divided into four different compartments.

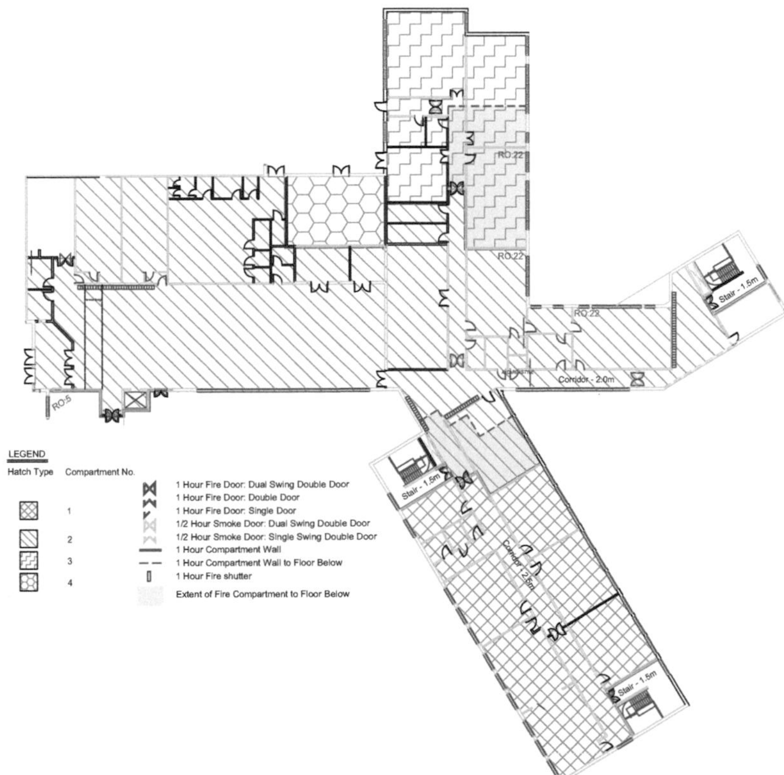

Figure 5.1 A typical compartment design for a tall building.

Compartment walls are used at the boundary of the compartments marked in blue color. Each compartment also has fire doors.

5.6.1 Key components in a compartment

As shown in Figure 5.2, a compartment is comprised of several key building elements, namely, compartment wall, compartment floor, fire door, and cavity barriers. These key building elements work together with other components to make an integrated compartment to contain the fire inside the compartment. All the key components are essential to keep the integrity of the compartment. For instance, cavity barriers will stop fire spread through the duct in floors or walls. Fire doors, compartment floors, and compartment walls also function as barriers in the event of fire. These building elements will be introduced in the following sections.

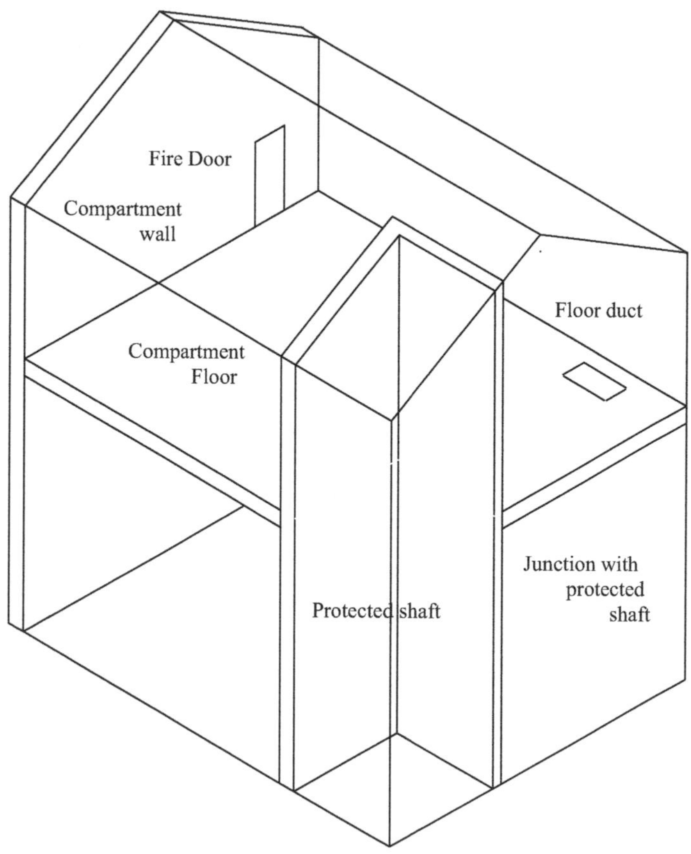

Figure 5.2 Typical compartment wall and compartment floor in a building.

5.6.1.1 Fire doors design

A fire door is one of the key components in a fire compartment. It is a specially manufactured door with a fire resistance rating to reduce the spread of fire or smoke between compartments. As shown in Figure 5.3, fire doors are made up of various materials, predominantly made from a solid timber frame in countries such as the UK, but they are sometimes covered by fire-resistant glass. Intumescent seal (see Figure 5.3) will be used around the edges of the door which is designed to expand when temperatures reach beyond 200°C to seal the gaps between the door and the frame. Cold smoke seals are embedded within the intumescent seal to stop the spread of the smoke as well. In addition, there are also some special requirements when manufacturing the door. These special requirements are demonstrated in Figure 5.3.

In the U.K., BS 8214 (2014) shows fire rating for a door can withstand; for example, an FD20 has been tested to withstand 20 min. The most common of the codes is FD30. The main categories of fire doors are FD30 and FD60 which offer 30 and 60 min fire protection, respectively.

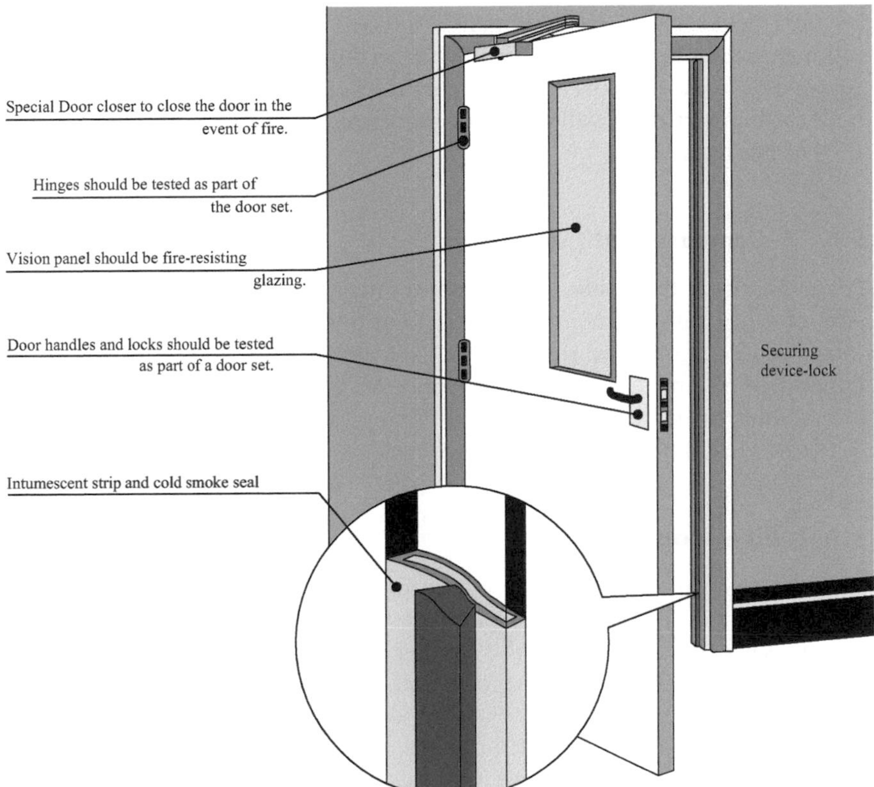
Special Door closer to close the door in the event of fire.
Hinges should be tested as part of the door set.
Vision panel should be fire-resisting glazing.
Door handles and locks should be tested as part of a door set.
Intumescent strip and cold smoke seal
Securing device-lock

Figure 5.3 Features of fire doors.

The Approved Document B: Appendix B: Fire Doors of Fire Safety states minimum requirements for fire doors depending on the position of the door in the building. For example, currently the requirement is that a fire door in a compartment wall separating two buildings should provide 60 min (FD60) fire protection, whereas a fire door affording access to an escape route should provide 30 min (FD30) fire protection.

The rating of fire doors is also determined through standard test procedures specified in BS 476-22 (1987) and BS EN 1634-1 (2014). Tests are made on complete fire door sets with the upper part of the door under a small positive pressure, to simulate the conditions likely to occur in a fire. It is tested from each side to establish its performance with either face exposed to fire conditions.

5.6.1.2 Compartment wall design

As shown in Figure 5.2, a compartment wall is a firewall with a fire-resistant rating to prevent the spread of fire for a prescribed period of time. Firewalls are built between or through buildings. Compartment walls form barriers between fire compartments and are required to provide a minimum degree of fire resistance as set out in Approved Document B1 (2019) Appendix A and Approved Document B2 (2019) Appendix A. The fire rating of a compartment wall is also tested through testing methods set out in BS 476 fire tests.

5.6.1.3 Compartment floor design

A compartment floor is horizontal component that forms part of the enclosure of a fire compartment. Compartment floors are required to provide a minimum degree of fire resistance as set out in Appendix A of Approved Document B2 and Appendix A of Approved Document B1. The fire rating of a compartment wall is also tested through testing methods set out in BS 476-13.1 (1983) fire tests.

5.6.2 Fire stop

In order to ensure the integrity of the fire compartment, it is critical that every opening or penetration through its walls, floor, and ceiling, from large openings such as doorways to the smaller service penetrations (cabling, piping, and ducts), is adequately protected to resist the passage of fire and smoke.

Approved Document B (2019) defines a fire stop as follows: "A seal provided to close an imperfection of fit or design tolerance between elements or components, to restrict the passage of fire and smoke."

Openings in compartment walls or floors (such as doors, windows, penetrating pipework, and ducts) and joints between compartment walls or floors are designed to have a similar fire resistance to the compartment walls or floors. They should be fire-stopped. In addition, openings for beams, joists, purlins and rafters, pipes, and ducts that pass through any part of a fire-separating element should be kept as few in number as possible, kept as small as practicable, and should be fire-stopped. Figure 5.4 shows some typical fire stops.

5.6.3 Cavity barrier

Fire may spread through cavities formed during the event of fire due to melting of thermal insulation materials. Cavity barriers are typically fire-stop materials which are fitted within building cavities—horizontally at each floor and vertically at each party wall. This assists with the integrity of fire compartmentation. In Approved Document B, Fire Safety, Volume 2 (2019),

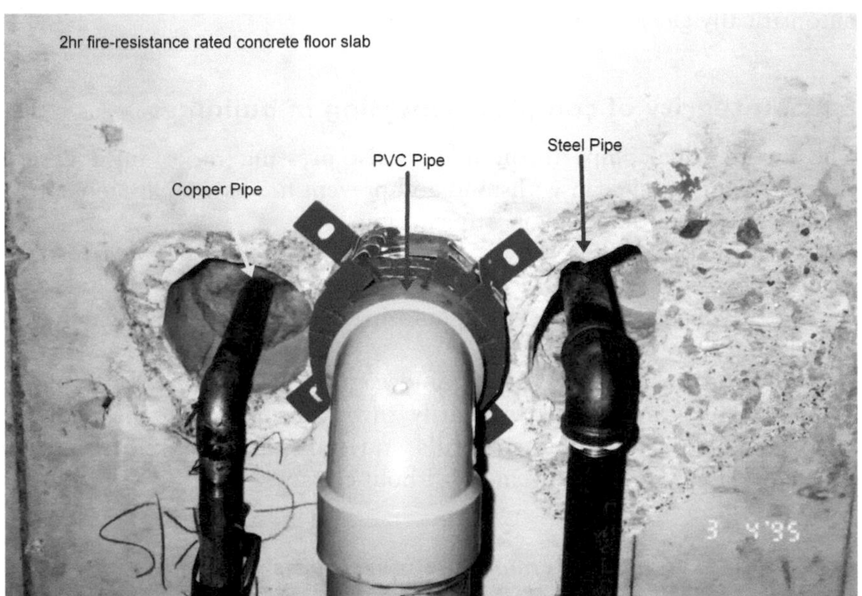

Northtown Casitas, North York (now Toronto), Ontario, Canada, 03.04.1995: Three firestopped pipe penetrations. The firestops on the copper pipe (left) and the steel pipe (right) are identical: Rockwool packing with 3M Fire Barrier Mortar on top. The pipe in the middle is plastic and it is fitted with a plastic pipe device using intumescent rubber 3M FS195 to choke off the melting plastic pipe in case of a fire. This view is from underneath, looking up at a concrete slab.

Figure 5.4 Fire stop. (This file is licensed under the Creative Commons Attribution-Share Alike 3.0 Unported license, https://en.wiktionary.org/wiki/firestop#/media/File:Nortown_casistas_3_pipe_firestops.jpg.)

a cavity barrier is defined as follows: "A construction within a cavity, other than a smoke curtain, to perform either of the following functions.

- Close a cavity to stop smoke or flame entering.
- Restrict the movement of smoke or flame within a cavity."

Effective cavity barriers are essential to restrict the spread of smoke or flames. If there is a fire, the intumescent material of the barrier will expand and seal off the gaps. Cavity barriers must be fitted tightly to rigid construction. The likelihood of failure of the fixings in the event of fire should be carefully considered. Services that run through barriers should also be made of appropriate fire-stop materials.

5.6.4 Fire damper

Fire dampers are installed in the ducts of heating, ventilation, and air-conditioning systems which penetrate compartment walls and floors, and will automatically close when they detect heat.

5.6.5 Integrity of compartmentation in buildings

The integrity of a compartment in fire is also pressing important. It ensures that the components can withstand and prevent fire as well as smoke from breaching the building's compartmentation.

As introduced in BRE (2005), *an integrity failure is deemed to occur when either collapse, sustained flaming, or gaps and fissures occur.* So far, limited guidance is available on maintaining the integrity of compartmentation during a fire.

Apart from using fire stoppable and cavity barriers, the integrity of a compartment is maintained primarily through controlling the allowable deflection of fire compartment walls and slabs. It must be designed to accommodate expected movement without collapse.

5.6.5.1 Measures to accommodate movements of compartment walls due to fire

In residential buildings, compartment walls generally go continuously down to the foundation, and so deflection of the walls will not be an issue. However, there is no guarantee that compartment walls will always be located in such an advantageous location.

The superseded BS 5950 Part 8 (2003) specify that *where a fire resisting wall is liable to be subjected to significant additional vertical load due to the increased vertical deflection of a steel beam in a fire, provision should be made to accommodate the vertical movement of the beam, or the wall should be designed to resist the additional vertical load in fire conditions. The anticipated vertical movement at mid-span of a vertically loaded steel beam in a fire should be taken as 1/100 of its span, unless a smaller value can be justified by an analytical assessment.*

SCI-P288 (2006) specifies that "compartment walls should, whenever possible, be located beneath and in line with beams." Therefore, for beams to pass over a compartment wall, either they should be protected or sufficient allowance for movement should be provided. BRE (2005) recommended that "a deflection allowance of span/30 should be provided in walls crossing the middle half of an unprotected beam. For walls crossing the end quarters of the beam, this allowance may be reduced linearly to zero at the supports."

For compartment walls made of lightweight plasterboard systems, the manufacturer can supply a range of standard details to accommodate movement from the floor above. The deflection allowance can be achieved through slide joints or deformable blankets or curtains.

5.6.5.2 Control movement of slab

As explained in Chapter 4, the deformation of the floor should not cause the failure of the compartmentation. Therefore, the limitation of deformation of composite slabs specified in it should be for the sake of integrity of the compartment. For other types of slabs, such as post-tensioning slab and flat slabs, similar requirements are needed.

5.7 EVACUATION ROUTE DESIGN

An effective evacuation route design is essential for a tall building in fire. The challenge of building evacuation increases with building height as it is difficult to ensure simultaneous evacuation of all occupants in buildings over 20 m in height. It may also cause additional risks when all of them evacuate via stairways. When designing an evacuation route plan for a tall building, the primary goal should be to provide an appropriate means to allow occupants to move to a place of safety.

To accomplish this goal, there are several evacuation methodologies:

1. Protection of escape routes
2. Fire-protected lift shaft and staircases
3. Refugee floor
4. Phased/progressive evacuation
5. A combination of these strategies.

When choosing an appropriate strategy, the required level of safety for the building occupants and the building performance objectives need to be considered.

5.7.1 Number of escapes routes and exits

The number of escape routes and exits is primarily decided based on the number of occupants. For a tall building, it is necessary to consider providing either of the following:

a. Separate escape routes from the areas of different use
b. Effective means to protect common escape routes.

5.7.2 Design of exits

The exit design is essential to enable occupants to quickly evacuate from the building. A tall building should provide a sufficient number of exits from a floor area. Due to large number of occupants in a tall buildings, they need at least two exits to evacuate from the area in the event of fire. In tall buildings, every part of each storey should have access to more than one stair. Areas in the dead end should also be provided access to an alternative stair. In designing the exit, the design of locations and travel distances of below exits and passage need to be considered.

Horizontal exit is the exit within one floor area.

Final exit refers to the threshold that separates "inside the building" and "outside the building." The liability of the design for evacuation from the building ends at this point.

Storey exit is an exit that gives direct access to a protected stairway, firefighting lobby, or external escape route.

Alternative exit every floor space shall be provided with at least two exits on the basis that if one exit is blocked by flame or smoke, the other exit can serve the function. Approved Document B (2019) requires that every storey with a floor level more than 11 m above ground level have an alternative means of escape.

5.7.3 Exit route

A protected exit route that leads from a storey exit to the final exit should be designed. Spaces that connect fire compartments, such as stairways and service shafts, need to be protected to restrict fire spread between the compartments. A refuge area can also be designed in between the storey exit and the final exit.

5.7.4 Travel distance

The travel distance to the nearest storey exit should be within the limits for routes as shown in Table 5.1. (The travel distance from every point in each storey should not exceed the distances given in it.) Where a sprinkle system

Table 5.1 Limitations on travel distance

Purpose group	Use of the premises or part of the premises		Maximum travel distance where travel is possible in: One direction only (m)	More than one direction (m)
2(a)	Residential (institutional)		9	18
2(b)	Residential (other):			
	a. In bedrooms(2)		9	18
	b. In bedroom corridors		9	35
	c. Elsewhere		18	35
3	Office		18	45
4	Shop and commercial		18	45
5	Assembly and recreation:			
	a. Buildings primarily for disabled people		9	18
	b. Areas with seating in rows		15	32
	c. Elsewhere		18	45
6	Industrial(3)	Normal hazard	25	45
		Higher hazard	12	25
7	Storage and other non-residential(3)	Normal hazard	25	45
		Higher hazard	12	25
2–7	Place of special fire hazard(4)		9(5)	18(5)
2–7	Plant room or roof-top plant			
	a. Distance within the room		9	35
	b. Escape route not in open air (overall travel distance)		18	45
	c. Escape route in open air (overall travel distance)		60	100

Reproduced from Table 2.1 of The Building Regulations 2010-Approved Document B, Volume 2 fire safety, Buildings other than dwellings, HM Government (2019), public domain, https://assets.publishing.service.gov.uk/government/uploads/system/uploads/attachment_data/file/832633/Approved_Document_B__fire_safety__volume_2_-_2019_edition.pdf.

is installed fully, the travel distance may be increased by 50% of the values specified in Table 5.1.

The ability of building occupants to escape from a building during a fire with travel distance be maintained safe is known as tenability. Tenability is assessed through parameters such as visibility, carbon monoxide levels, and temperature.

Table 5.2 Minimum width of staircase (National Code of India, 2005)

Residential buildings (dwellings)	1.0 m
Residential hotel buildings	1.5 m
Assembly buildings like auditorium, theaters, and cinemas	2.0 m
Educational buildings up to 30 m in height	1.5 m
Institutional buildings like hospitals	2.0 m

5.7.5 Staircases and elevators

Staircases and elevators can also be used as evacuation routes. The number of staircases and elevators shall be decided in conjunction with the number of exits. As stipulated in Approved Document B (2019), single staircase should only be allowed in residential or office buildings with height less than 12 m. Therefore, for tall buildings, more than two staircases and elevators shall be designed.

To facilitate fast movement of evacuees, most of the codes specify the minimum width of the staircase. National Code of India (2005) specifies the minimum widths of staircases as shown in Table 5.2.

5.7.5.1 Protected staircases and elevators

Using protected elevators and staircases is an essential evacuation strategy in the design of tall buildings due to the height of the building. In addition, protected staircases and lifts also assist firefighters with operations and rescues. Every internal escape stair should be a protected stairway (within a fire-resisting enclosure). In certain buildings, a protected shaft or firefighting shaft is also designed.

There are a number of design considerations to consider for protected elevators:

- Evacuation strategy
- Safety and reliability of the elevators
- Coordination of elevator controls and building safety systems
- Communication devices in the elevator.

Figure 5.5 shows three common examples of protected lift shaft.

5.7.5.2 Fire lift lobby

Every lift lobby should have access, without any obstruction and lockable door, to an exit route. Such access should be available at all times to any person who may come out from a lift car to the lift lobby. The provision of a direct intercom link connecting a lift lobby with the management office of the building will be accepted as an adequate alternative.

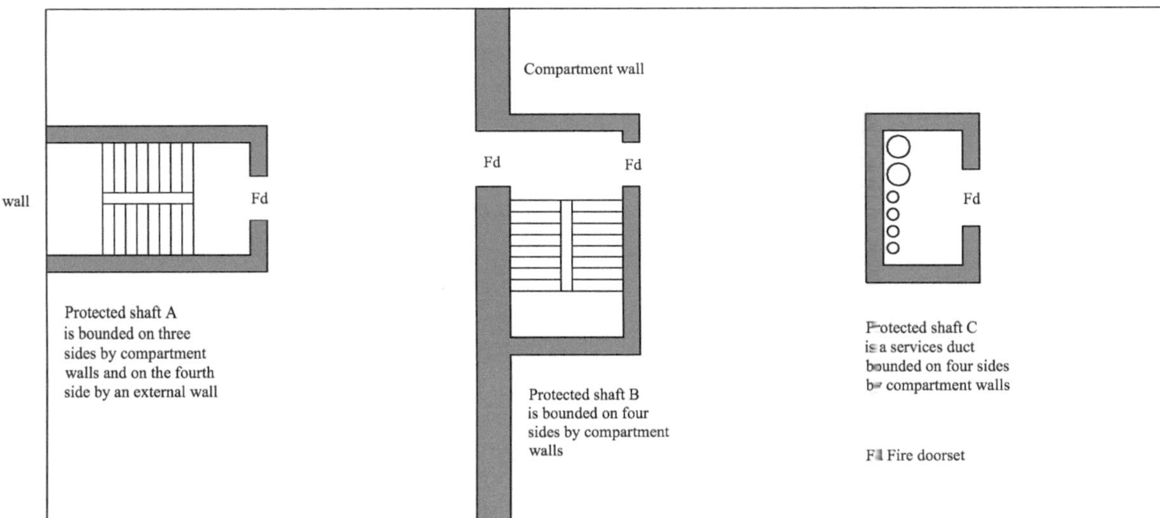
The diagram shows three common examples which illustrate the principles. The elements enclosing the shaft(unless formed by adjacent external walls) are compartment walls and floors
External wall
Compartment wall
Fd
Fd
Fd
Fd
Protected shaft A
is bounded on three
sides by compartment
walls and on the fourth
side by an external wall
Protected shaft B
is bounded on four
sides by compartment
walls
Protected shaft C
is a services duct
bounded on four sides
by compartment walls
Fd Fire doorset
The shaft structure (including any openings) should meet the relevant provisions for:
compartment walls (see paragraphs 8.15 to 8.31),external walls (see sections 12 and 13 and Diagram 3.3

Figure 5.5 Protected lift shaft. (Reproduced from Diagram 8.3 of The Building Regulations 2010-Approved Document B, Volume 2 fire safety, Buildings other than dwellings, HM Government (2019), in public domain, https://assets.publishing.service.gov.uk/government/uploads/system/uploads/attachment_data/file/832633/Approved_Document_B__fire_safety__volume_2_-_2019_edition.pdf.)

5.7.5.3 *External escape stairs*

Another escape route may be an external escape staircase, provided that there is at least one internal escape staircase from every part of each storey. However, the route is not intended for use by the public.

5.7.6 Phased/progressive evacuation

Most tall buildings will adopt a phased evacuation regime due to its height. In phased evacuation approach, only people in the immediate vicinity of the fire are evacuated, and thus it allows people in direct danger to make most efficient use of the egress provisions available to them. Typically, those in most danger are evacuated first while surrounding areas (zones) are put on "alert." Those occupants within the zone being evacuated are given an appropriate warning signal, while those outside the evacuation zone are either notified of a developing incident and told to remain in place and await further instruction, or are given no warning at all and continue their normal day activities unaware of any incident.

5.7.7 Refuge

For a tall building, it is difficult to evacuate the occupants from a higher floor down to the ground in a short time. Therefore, it is a common practice to design a refugee shelter or an entire refugee floor for the occupants to temporarily stay in the event of fire at high levels of the buildings. Refuges offer relatively safe areas for people to wait for a short period. Occupants of the evacuation zone are evacuated to the refuges from the fire location (as opposed to evacuating directly to outside). Occupants can either remain in this safe place or, if threatened further, be relocated to an alternative safer area within the building.

Approved Document B (2019) specifies that refuges should meet the following conditions:

- Refuges should be provided on every storey (except the ones consisting only of plant rooms) of each protected stairway providing an exit from that storey.
- Refuges do not need to be located within the stair enclosure but should enable direct access to the stair.
- A single refuge may be occupied by more than one person during the evacuation procedure.

Rather than building a separate refuge, a compartment, protected lobby, protected corridor, or protected stairway, an area in the open air, such as a flat roof, balcony, podium, or similar place, can be used as a refuge, given that they do not reduce the width of the escape route or obstruct the flow of people escaping.

They should also have the following features:

- Protected (or remote) from any fire risk
- Have their own means of escape.

Refuges should be provided with an emergency voice communication (EVC) system complying with BS 5839-6 (2019).

5.7.8 Clear sign for evacuation

To enable fast evacuation, clear signs are also needed. For instance, Refuges and evacuation lifts should be clearly identified. In protected lobbies and protected stairways, there should be a blue mandatory sign worded "'Refuge—keep clear" in addition to fire safety signs.

5.7.9 Computational models for evacuation simulation

The evacuation design is very challenging due to the different types of behaviors of humans with different types of personalities, and different types of building configurations. It is crucial to manage evacuation upstream in a situation of emergency with consideration of panic made to evacuees. In recent years, several fire evacuation models have been proposed for a fire safety engineer to assist the fire engineer in evacuation route design.

Agent-Based Model (Kasereka et al., 2018) enables the modeling and simulation of evacuation of people from a building on fire. It is based on four parameters—evacuees, fire, smoke, and alarm—to allow the evaluation, considering different scenarios for each parameter and the interaction of these parameters.

Evac (Valentin, 2013) is a free evacuation system to simulate evacuation and fire propagation. The decisions of individuals are simulated based primarily on the preference of the exit category, optimization of expected time up to the exit, propagation simulation of incompressible fluid in order to determine the path to follow, and social and instinctive behaviors of each individual (fear of fire and smoke).

Fire Dynamics Simulator (FDS) is a large-eddy simulation (LES) code for low-speed flows, with an emphasis on smoke and heat transport from fires. FDS-evac can be used to test the human evacuation in case of fire initiation, and this is used to help determine necessary time for evacuation.

5.8 EMERGENCY VEHICLE AND FIREFIGHTER ACCESS

To enable swift access of firefighters, the building layout should be simple to enable firefighters to locate an area quickly. However, this is not always possible; especially for commercial buildings, the layout is increasingly complicated.

5.8.1 Equipment for firefighting

The building should be equipped with fire department connections, fire command center, water pump room, hose valves, fire hydrants, etc.

5.8.2 Firefighting lift, lobby, shaft, and stair

To facilitate the fire and rescue service, some special facility can be designed such as

1. A direct control of a fire protected lift by fire service.
2. A protected lobby providing access.
3. A protected enclosure containing a firefighting stair, firefighting lobbies, and a firefighting lift, together with its machine room.
4. A protected stairway communicating with the accommodation area only through a firefighting lobby.

5.9 RESISTING FIRE SPREAD THROUGH BUILDING ENVELOPE

As the lessons learned from Grenfell Tower show, besides containing or slowing down the fire speed internally, building envelope should also resist the spread of fire. The external envelope of a building should not contribute to fire spread from one part of a building to another part.

Following the Grenfell Tower fire, a decision was taken to ban combustible materials in the building envelope over 18 m in height. This will help prevent the spread of fire from one compartment to another through the outside of a high-rise building. The following change to Approved Document B (2019) specified that

> In a building with a storey 18m or more in height any insulation product, filler material (such as the core materials of metal composite panels, sandwich panels and window spandrel panels but not including gaskets, sealants and similar) etc. used in the construction of an external wall should be class A2-s3, d2 or better. This restriction does not apply to masonry cavity wall construction which complies with Diagram 9.2 in Section 9.

5.9.1 Resisting fire spread over external walls

The external surfaces (i.e. outermost external material) of external walls should comply with the provisions in Approved Document B (2019) (Table 5.3). The provisions in the table apply to each wall individually in relation to its proximity to the relevant boundary.

Table 5.3 Reaction of fire perfromance of external walls

Building type	*Building height*	*Less than 1,000 mm from the relevant boundary*	*1,000 mm or more from the relevant boundary*
'Relevant buildings' as defined in regulation 7(4) (see paragraph 12.11)		Class A2-s1, d0(1) or better	Class A2-s1, d0(1) or better
Assembly and recreation	More than 18 m	Class B-s3, d2(2) or better	From ground level to 18 m: class Cs3, d2(3) or better From 18 m in height and above: class B-s3, d2(2) or better
	18 m or less	Class B-s3, d2(2) or better	Up to 10 m above ground level: class C-s3, d2(3) or better Up to 10 m above a roof or any part of the building to which the public have access: class C-s3, d2(3) or better(4) From 10 m in height and above: no minimum performance
Any other building	More than 18 m	Class B-s3, d2(2) or better	From ground level to 18 m: class Cs3, d2(3) or better From 18 m in height and above: class B-s3, d2(2) or better
	18 m or less	Class B-s3, d2(2) or better	No provisions

Reproduced, Table 12.1 of British HM Government Approved Document B (2019), Fire safety, in public domain, https://assets.publishing.service.gov.uk/government/uploads/system/uploads/attachment_data/file/832633/Approved_Document_B__fire_safety__volume_2_-_2019_edition.pdf.

Notes: In addition to the requirements within this table, buildings with a top-occupied storey above 18 m should also meet the provisions of paragraph 12.6. In all cases, the advice in paragraph 12.4 should be followed.

1. The restrictions for these buildings apply to all the materials used in the external wall and specified attachments (see paragraphs 12.10–12.13 for further guidance).
2. Profiled or flat steel sheet at least 0.5 mm thick with an organic coating of no more than 0.2 mm thickness is also acceptable.
3. Timber cladding at least 9 mm thick is also acceptable.
4. 10 m is measured from the top surface of the roof.

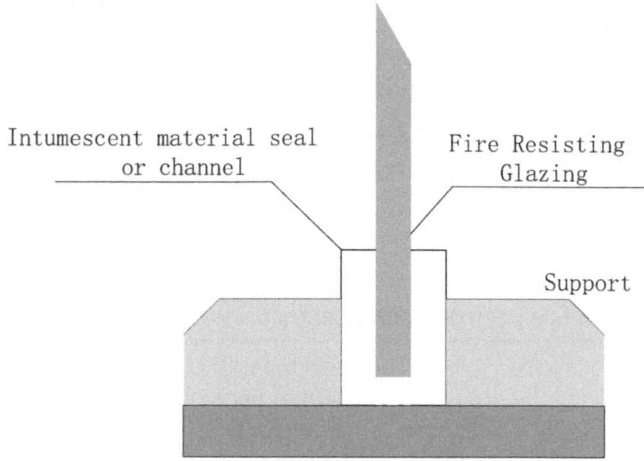

Figure 5.6 Schematic diagram of fire resistance glazing.

This specification restricts the risk of internal fire to spread through the external wall as well as possible ignition by an external source to the outside surface of the building, and spread of fire over the outside surface should also be restricted.

However, the most widely used curtain wall in tall buildings introduced flammable materials into both the wall linings and external cladding. The methodologies used to define the fire resistance of these systems are ambiguous, i.e. they do not take into account deformations possible with evolving fires.

5.9.2 Fire-resisting design for glazing

For fire-resisting design of glazing, fire-resistant glasses can be used. The size of the glass and the method of its retention are important factors that influence its integrity. It provides protection by impregnation of a surface coating or a surface covering of non-combustible materials or to fit a fire-resistant glass secured using a fire-resistant glazing system. This will hold the glass firmly in place during normal use, but in the event of fire, it allows the intumescent material to expand, thereby securing and insulating the glass and protecting the surrounding timber. Figure 5.6 shows a typical fire resistance glazing.

5.10 FIRE DETECTION, ALARM, AND COMMUNICATION SYSTEM

An accurate alert to building occupants during emergencies increases their chance to safety. Therefore, fire alarm and communication systems are important facilities in the warning system. Very tall buildings employ voice communication systems that are integrated into the fire alarm system.

Survivability of the fire alarm system is another import factor to consider in fire alarm system design. It should be considered while designing fire alarm system for tall buildings so that an event of fire in an evacuation zone will not impair the voice communication outside the zone. Some of the design considerations to achieve survivability may include (1) protection of control equipment from fire, (2) protection of circuits, (3) configuration of circuits, and (4) shielding of panels.

5.10.1 Central fire alarm system

It is essential for large buildings to be equipped with a fire alarm system. Fire alarm system connected with smoke and heat detectors will alert people through visual and audio appliances. When smoke or fire is detected, the system will automatically trigger the fire control system Figure 5.7 shows a typical fire alarm device.

5.10.2 Smoke detections

A smoke detector is a device that senses smoke, typically as an indicator of fire. It is an important tool in fire safety design of tall buildings. Smoke detector usually is directly connected or powered by a central fire alarm system to ensure that notification can be passed to occupants in the building. All lift lobbies shall be provided with smoke detectors. Smoke detectors can be divided into two types: photoelectric smoke detectors and ionization smoke detectors.

Figure 5.7 A typical fire alarm device.

5.10.3 Smoke control

Controlling the spread of smoke allows a safe means of escape. It is more complicated in tall buildings. Stack effect occurs when a tall building experiences a pressure difference throughout its height as a result of temperature difference between the outdoor and indoor temperatures. It can also cause smoke from a building fire to spread throughout the building if not controlled. That is why tall buildings often employ smoke management systems that either vent, exhaust, or limit the spread of smoke.

Other concerns in tall buildings include the air movement created by the piston effect of elevators and the effects of wind. Air movement caused by elevator cars ascending and descending in a shaft and the effects of wind can result in smoke movement in tall buildings.

Due to the above complexities of smoke spread, an effective smoke control is more difficult to achieve. The possible solutions are as follows:

a. Smoke barrier walls and floors
b. Stairway pressurization systems, pressurized zoned smoke control provided by the air-handling equipment
c. Smoke dampers.
d. Venting of smoke and heat: exhaust ventilation, natural and mechanical ventilation
e. Smoke extraction system
f. Smoke dilution: post-fire smoke clearance and enhance system
g. High-power fans.

5.11 FIRE AND SMOKE SUPPRESSION SYSTEM

In a tall building, fire suppression systems are quite often used due to its height. This kind of system prevents ignition. Any fire that burns with sufficient availability of heat, oxygen, and fuel can be controlled by a suppression system. They can be categorized as follows:

Reducing oxygen levels: Fire suppressants such as INERGEN, IG55, and IG541 are designed to reduce oxygen levels to below 15% but higher than 12%. If the oxygen level is below 15%, combustion is not possible.

Reducing heat: These fire suppressants are designed to reduce heat. Due to the nature of the fire suppression technique, these refrigerants are required only in small concentrations. This means less cylinders and lower pressures. Oxygen levels are also not significantly reduced, which makes them safe to use within occupied spaces.

Water-based systems: Generally these systems are not suitable for rooms with equipment, such as a computer rooms. Water mist fire suppression system is particularly good for extinguishing or suppressing fire that burns at high temperatures.

Water sprinkler fire suppression systems: It stops the spread of fire by spreading water. The problem with sprinkler systems is that they can sometimes do more damage than the fire itself. One of the famous examples is the incident of WTC7 introduced in Chapter 2. The collapse of the WTC7 is partially because of the malfunctioning of the water supply to the sprinkler system.

Foam fire-extinguishing systems: Water mixed with high-expansion foam solution such as AFFF allows the foam to rapidly smother the area with foam. This is not suitable for most common fire suppression applications but particularly good for large open spaces, where spillage of fuel or other types of combustible liquids are present.

Carbon dioxide suppression: A CO_2 fire suppression system eliminates the oxygen to suppress the fire. When the suppression system detects smoke or fire, it then releases the CO_2 agent into the space it is protecting. The CO_2 level in the space quickly increases and thus the oxygen level quickly drops causing the fire to be suppressed or extinguished. As any sensitive equipment that is in the protected space is not damaged by the CO_2, which reduces downtime and costs.

5.12 COMPARISON FOR FIRE PROTECTION SYSTEM FOR TALL BUILDINGS ACROSS THE WORLD

The fire protection systems vary between different types of tall buildings. Table 5.4 summarizes the fire protection systems of major tall buildings across the world. As it has been discussed in Chapter 2, the failure of columns triggered the collapse of both WTC1 and WTC7. Column plays an important role in maintaining the stability of the building during fire. It can be noticed that in Table 5.4, all the buildings provide 3 h fire protection for columns, to prevent early collapse or local collapse of the building. To enable firefighter to get swift access to the building occupants, all the buildings provided protected lifts for firefighters.

5.13 CASE STUDY OF FIRE SAFETY DEIGN FOR BURJ KHALIFA

Figure 5.8 shows the tallest building in the world Burj Khalifa. As it is the tallest building in the world, its fire safety design becomes a challenging task. It is 828.8 m tall with 160 floors, and the maximum occupant capacity is 35,000 people.

Table 5.4 Fire protection systems of tall buildings across the world

Fire protection	*Burj Dubai*	*Jin Mao Building*	*Petronas Towers*	*John Hancock Center*	*Sears Tower*
Fire compartment size	1 per floor	2,000 m² in office and parking areas	2,000 m² in office areas	1 per floor	1 per floor
Fire resistance	Columns: 3 h Floors: 2 h	Columns: 3 h Floors: 2 h	Columns: 3 h Floors: 2 h	Columns: 3 h Floors: 1 ½ h	Columns: 3 h Floors: 3 h
Quick response sprinklers	Yes	No	No	No	No
Primary fire water supply and duration	Basement tank with secondary tanks distributed vertically throughout the tower with pumps	Basement tank connected to domestic water tank, two secondary water tanks at level 51 and penthouse with pumps	Basement tanks connected to domestic water mains with fire pumps	Dual source domestic water mains	Dual domestic water mains with water tanks and fire pumps distributed vertically in building.
	2 h internal, 4 h total	Unknown	1 h for sprinklers	1/2 h	
Firefighting lift	2 up to level 111 1 for level 112 through 160	2	2	1	1

(*Continued*)

Table 5.4 (Continued) Fire protection systems of tall buildings across the world

Fire protection	*Burj Dubai*	*Jin Mao Building*	*Petronas Towers*	*John Hancock Center*	*Sears Tower*
Smoke-controlled exit stairs	Pressurized	Pressurized stairs and vestibules	Pressurized	Vestibule with naturally ventilated smoke shaft	Vestibule with naturally ventilated smoke shaft
Floor smoke control	Provided	Provided	Provided	Provided	Provided
Fire command center	Primary and secondary provided	Primary and secondary provided	Provided	Provided	Provided
Emergency P.A. system	Provided	Provided	Provided	Provided	Provided
Refuge areas	Levels 42, 75, 111, and 138	Levels 15, 30, 58–85 (hotel levels) and penthouse level 2			
Smoke compartment size	1 per floor	1 per floor	2,000 m^2 in office areas	1 per floor	1 per hour

Source: Evenson and Vanney (2008).

Figure 5.8 Burj Khalifa. (Photo taken by the author.)

As shown in Figure 5.9, it uses a so-called buttressed core structural stability system, with a major hexagonal core wall in the center, buttressed by three wings of shear walls. Each wing is further braced by another group of short shear walls.

When evaluating the appropriate degree of fire protection for the building, the design team (Evenson and Vanney, 2008) applied three fundamental life safety concepts:

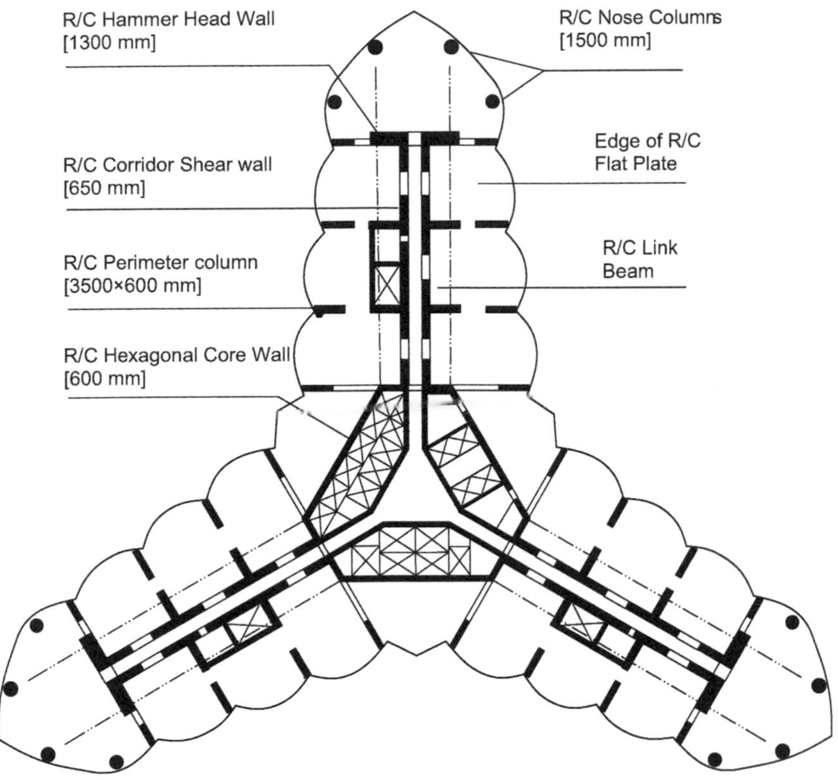

Figure 5.9 Plan layout of Burj Khalifa (Evenson and Vanney, 2008).

1. Control the fire and its effects
2. Accommodate occupant relocation/evacuation
3. Facilitate firefighting operations.

5.13.1 Evacuation and refuge

The phased/progressive evacuation is designed for Burj Khalifa. Along the 160 floors, there are some mechanical floors which are serving as the area of refuge, and they are at levels 42, 75, 111, and 138. These refuges are separated from the remainder of the building by 2 h fire resistance rated construction and are pressurized to minimize the migration of smoke into the compartment. In addition, refuges are connected to various tower stairwells, allowing the occupants to evacuate their specific location and move to the areas of refuge and, if directed, proceed down the tower through multiple protected stairwells.

As it has been introduced, it is difficult to evacuate all the occupants to the ground when building height increases. To tackle this problem, an innovative elevator-assisted evacuation strategy is used. It uses specially designed lifts to transport occupants from predetermined levels throughout the tower. This specially designed lift features in emergency backup power and fire resistance, to ensure that it can be used in the event of fire.

5.13.2 Firefight access

There are two firefighter's lifts: one from levels 2 to 111 and the other from levels 112 to 160.

5.13.3 Staircase and elevator

Pressurized smoke-protected exit staircase was designed. Apart from this, specially designed high-power fans were installed in the staircase for driving off the smoke, to clear the evacuation route of smoke and flame in exit route for the evacuees.

5.13.4 Alarm and warning system

Voice communication system of the central fire alarm is used to alert the occupants in various building zones by conveying select messages.

5.13.5 Fire suppression

A combination of sprinkler and standpipe system, external hydrant system, preaction system, and foam system is used in this project.

5.13.6 Special water supply and pumping system

As introduced in Chapter 2, one of the reasons that caused the collapse of WTC7 is the failure of water supply system. Therefore, water supply system is important in both fire suppression and firefighting. In Burj Khalifa, a basement tank and several secondary tanks distributed vertically throughout the tower with pumps are used for water supply. The primary storage pump was installed at level B2 with 2 h capacity; the water is pumped to level 74's tank with 1 h capacity as well as flow to a gravity-filled tank at level 40 with 1 h capacity. The water is continually pumped to level 137 in a tank with 1 h capacity and flow to a gravity-filled tank at level 109 with 1 h capacity. This water supply system ensures the water supply at different levels of the building, as well as avoids relying on single water storage.

5.14 CASE STUDY: STRUCTURAL FIRE DESIGN OF THE SHARD

As shown in Figure 5.10, the Shard is one of the few tallest buildings in Europe for fire safety design. WSP Global is the structural engineering design consultant of this project. The author was luckily involved in the

Figure 5.10 The Shard. (The photo is under CC BY-SA 4.0 license agreement, https://upload.wikimedia.org/wikipedia/commons/0/07/The_Shard_from_the_Sky_Garden_2015.jpg.)

structural fire design of the tallest building in Western Europe, the Shard. It is my pleasure to share some of the design experience of structural fire design for this tallest building in the U.K.

5.14.1 Introduction of the project

As shown in Figure 5.10, the Shard is a 310 m tall building with 95 floors. It is a mixed-use high-rise building designed to accommodate retail, offices, viewing galleries, hotel, and apartments together with associated plant rooms, parking, loading bays, and other ancillary facilities.

For a building of this size and nature, a range of approaches were considered to ensure that the fire protection specified meets the requirements of all stakeholders to the project while ensuring an added level of confidence that the structural fire performance of the building is acceptable under the expected range of fire scenarios. Therefore, a mixed design approaches: prescriptive design approach for the concrete areas of the structure (deemed to satisfy rules) and performance-based approach in other areas of the structure are used to generate a greater degree of confidence in the proposed design solution.

5.14.2 Structural system

As shown in Figure 5.11, the Shard is primarily a steel-framed building using mixed post-tensioning slabs and composite metal decking as the major floor system. As introduced in Fu (2018), the major lateral stability system for the Shard is concrete Core+Outrigger system. Transfer trusses are also used in certain floors due to the change of the structural layout.

The following key design tasks in structural fire design have been identified by the design team:

1. Fire resistance design for the main structural steelwork between open office floors from levels 2 to 40 using structural fire design and analysis approach.
2. Fire resistance design for external columns at the bottom of the structure using external flaming calculations.
3. Fire resistance design for main transfer structures under the effects of severe and highly localized fires.
4. A qualitative assessment of the likely performance of PT flat-slab construction in fire.

5.14.3 Determine the worst-case fire scenarios

While designing the Shard, both deterministic and probabilistic approaches have been adopted to determine the worst-case fire scenarios for different

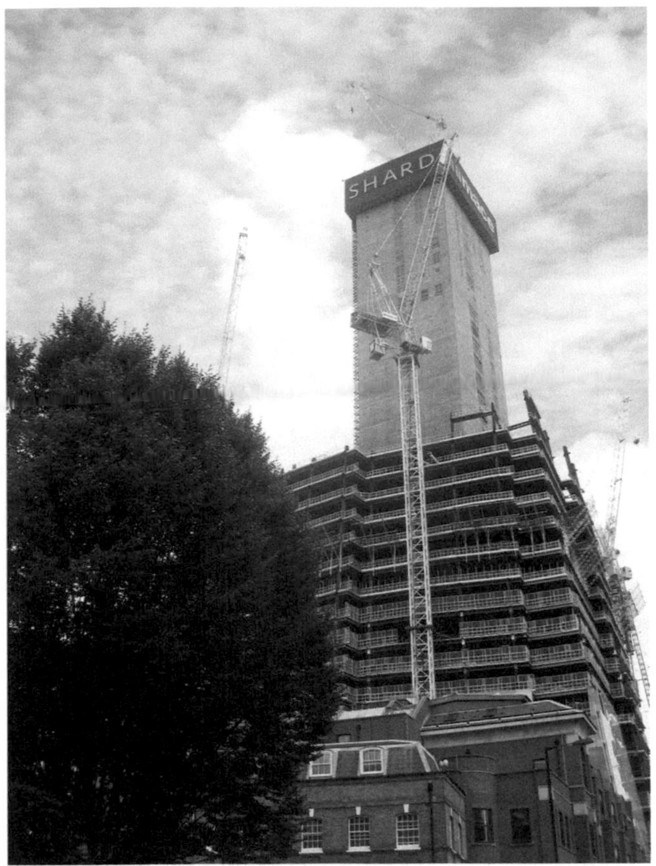

Figure 5.11 The structural system of the Shard. (This file is licensed under the Creative Commons Attribution 2.0 Generic license, https://commons.wikimedia.org/wiki/File:Shard_London_Bridge_July_2010.jpg.)

floors according to the usage of the floor area. In the deterministic approach, the sensitivity of the key input variables of fire loading and ventilation conditions is assessed for a sample of structural element types in a number of compartments. A Monte Carlo simulation is used for the probabilistic approach. The opening factor and fire density are sampled using the Monte Carlo simulation to determine the worst-case fire scenarios.

5.14.4 Design for fire resistance

The fire resistance scheme is determined through thermal–mechanical analysis using commercial software Abaqus®. As shown in Figure 5.12, a 3D model was built in Abaqus® by the author. It shows the temperature

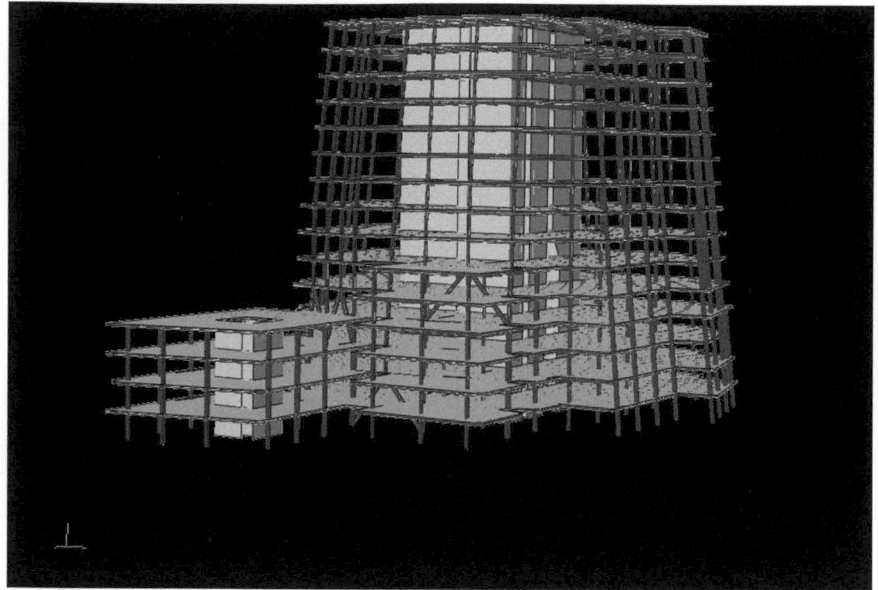

Figure 5.12 3D finite element model of Shard using Abaqus®. (Abaqus® screenshot reprinted with permission from Dassault Systèmes.)

distribution for the office levels of the building. Other results such as deflection, stress, strain, and member forces can also be obtained from the model.

An optimized fire protection scheme was proposed. All the primary beams are fire protected, rather than protecting all the secondary beams. Therefore, the cost for fire protection has been dramatically reduced. Based on the requirement of the code, various fire resistance rating for the structural members (from 60 to 120 min) were used. Acceptance of the suitability of each system will be assessed based upon its structural performance under the range of fires studied and the consequences of failure of the individual element.

5.15 STRUCTURAL FRAMING AND STRUCTURAL SYSTEM

From the above introduction, it can be seen that the structural framing and the structural system also play significant roles in fire safety design for tall buildings. As shown in Figure 5.13, most of the tall buildings use core as a major stability system.

One of the reasons of the collapse of Twin Towers is the use of steel core which has poorer fire resistance compared to concrete. As a result of the 9/11 attack, more and more design engineers focus on how to design a tall building to be able to resist a similar attack. Apart from this reason, as

Figure 5.13 A concrete core. (Photo taken by the author.)

discussed, core is a key part of evacuation route when fire happens, as all the staircases and lifts are built inside core. Protected lift is also a major access for firefighters to get into the building, and it should be kept flame and smoke free. Therefore, concrete core would be a good option.

In terms of the plan layout, to ensure travel distance to the exit within the limit shown in Table 5.1, except main core at center, Burj Khalifa also has an extra staircase at each wing as shown in Figure 5.9. In the podium of the Shard, a separate core was also built as shown in Figure 5.13.

REFERENCES

BRE report (2005), *The Integrity of Compartmentation in Buildings during a Fire*, Building Research Establishment Ltd.

BS 476-22 (1987), Fire tests on building materials and structures. Methods for determination of the fire resistance of non-loadbearing elements of construction.

BS 476-23 (1987), Fire tests on building materials and structures. Methods for the determination of the contribution of components to the fire resistance of a structure.

BS 5950–8 (2003), Structural use of steelwork in building. Code of practice for fire resistant design, Status: Superseded, Withdrawn Published: November 2003. Replaced By: BS EN 1993-1-2:2005.

BS 8214 (2014), Code of practice for fire door assemblies.

BS 5839-6 (2019), Fire detection and fire alarm systems for buildings. Code of practice for the design, installation, commissioning and maintenance of fire detection and fire alarm systems in domestic premises.

BS EN 1634-1 (2014), Fire resistance and smoke control tests for door, shutter and openable window assemblies and elements of building hardware. Fire resistance tests for doors, shutters and openable windows which is an alternative for BS 476-22: 1987.

Evenson, J. M., Vanney, A. F. (2008), Burj Dubai: Life Safety and Crisis Response Planning Enhancements, *CTBUH 8th World Congress 2008*.

Fu, F. (2018), *Design and Analysis of Tall and Complex Structures*, Elsevier. ISBN 978-0-08-101018-1.

Hadjisophocleous, G., Benichou, N. (1999), Performance criteria used in fire safety design. *Automation in Construction*, 8, pp. 489–501. (Morgan Hurley Fire Protection Engineering.)

HM Government (2019), The Building Regulations 2010-Approved Document B, Volume 1 fire safety, Dwellings', HM Government.

HM Government (2019), The Building Regulations 2010-Approved Document B, Volume 2 fire safety, Buildings other than dwellings', HM Government.

International Code Council/SFPE (2013), *Engineering Guide: Fire Safety for Very Tall Buildings*, Country Club Hills, IL.

Kasereka, S., Kasoro, N., Kyamakya, K., Goufo, E.-F. D., Chokki, A. P., Yengo, M. V. (2018), Agent-based modelling and Simulation for evacuation of people from a building in case of fire. *Procedia Computer Science*, 130, pp. 10–17.

National Building Code of India Part 4 (2005), Fire and Life Safety-2005 (Second Revision of SP 7-Part 4).

SCI Publication P288 (2006), *Fire Safety Design: A New Approach to Multi-Storey Steel-Framed Buildings*, 2nd edition, Steel Construction Institute. Ascot, ISBN 1859421202

SFPE (2011), SFPE Standard S.01 2011, Engineering Standards on Calculating Fire Exposures to Structures, Gaithersburg, MD.

SFPE (2015), SFPE Standard S.02 2015, SFPE Engineering Standard on Calculation Methods to Predict the Thermal Performance of Structural and Fire Resistive Assemblies, Gaithersburg, MD.

Valentin, J., (2013), Simulation du comportement humain en situation d'évacuation de bâtiment en feu. PhD Thesis, Univ. De Pau, France.

Chapter 6

Fire analysis and modeling

6.1 INTRODUCTION

As it has been introduced in the preceding chapters, fire analysis dominates the whole fire safety design process. For instance, in structural fire design, determining compartment temperature and thermal response of individual building elements requires an accurate structural fire analysis procedure to be adopted. When determining thermal behavior, material nonlinearities and degradation, as well as various boundary conditions, must be considered in most cases, making numerical modeling essential in solving these problems.

Therefore, in this chapter, various theoretical and numerical methods for fire analysis will be introduced. It starts with the method to determinate the compartment fire including a detailed introduction of zone model and CFD model. It is followed by the introduction of solving thermal response of structural members such as heat transfer analysis and thermal-mechanical analysis processes. In addition, the probabilistic method in fire safety analysis will be introduced. In the final part of this chapter, various numerical modeling software for fire analysis will be explained.

6.2 DETERMINING COMPARTMENT FIRE

As it has been introduced in the preceding chapters, determining the compartment fire temperature is the first step of fire safety design. Two categories of natural fire models are specified in Eurocode 1 (EN 1991-1-2, 2002): simplified fire models and advanced fire models (can model gas properties, mass exchange, and energy exchange).

6.2.1 Simplified models from Eurocode

Simplified model uses the formulas from Eurocode 1 (EN 1991-1-2, 2002) to determine the compartment fire. Eurocode 1 gives two major types of fire models: one for ordinary compartment fire and the other for localized fire.

6.2.1.1 Compartment fires

Compartment temperatures (also called gas or atmosphere temperature) should be determined on the basis of physical parameters (the two influential factors are fire load density and the ventilation conditions). As introduced in Chapter 3, two major fire temperature curves are used in the current design practice: parametric fire temperature curve (Eurocode 1, EN 1991-1-2, 2002) and standard fire temperature curve (BS476: part20, 1987). Standard fire curve is primarily used for fire resistance rating test, and the temperature keeps rising without considering the cooling stage of the fire. Therefore, it is not a representation of the real fire. However, for historical reason, most engineers are still using it in fire safety design. As it is widely believed, it gives a more conservative design. However, the research from Cardington tests (BRE, 2004) discovered that most of the structural members actually failed in the cooling stage, due to the contraction of the structural members. Therefore, parametric fire temperature curve considers both the heating phase and the cooling phase to give a closer representation of the real fire, and so is recommended to be used in the design.

6.2.1.2 Localized fires

As introduced in the preceding chapters, localized fire is prone to occur in tall buildings mostly due to their large open plan. A method for the calculation of thermal actions from localized fires is given in annex C of EN 1991-1-2: Eurocode 1, which has been introduced in Chapter 3.

6.2.2 Advanced models

Advanced models are primarily implemented through commercial software. The theory underpins their analysis, which will be introduced here.

6.2.2.1 Zone models

Zone models determine the compartment fire through dividing the fire compartment into certain zones with different temperatures. In each zone, it is presumed that atmosphere temperature, density, and internal energy and pressure are uniform. By solving the governing equation of conservation of mass and energy, the thermal action can be solved. This model requires input parameters such as fuel loads, compartment size, boundary condition, and thermal properties (Fu, 2016a). There are two models commonly used: one is two-zone model, and the other is one-zone model.

6.2.2.1.1 Two-zone model

Two-zone model considers non-uniform distribution of temperature inside the compartment but a uniform distribution within each layer. In each zone, uniform temperature is presumed (as shown in Figure 6.1). It can be used for pre-flashover localized fire modeling. The Eurocode 1 (EN 1991-1-2, 2002) model for localized fire is one example of two-zone model. It divides the fire compartment into two separate zones. The upper layer represents the accumulation of smoke, and the bottom layer represents fire heat.

In Figure 6.1, the variables considered to describe the gas in a compartment are as follows:

Q is the energy terms,

ρ_{int} is the internal pressure,

m_U and m_L are the masses of the gas of, respectively, the upper and lower layers,

T_U and T_L are the gas temperatures of, respectively, the upper and lower layers,

V_U and V_L are the compartment volumes of, respectively, the upper and lower layers,

E_U and E_L are the internal energies of, respectively, the upper and lower layers,

ρ_U and ρ_L are the gas densities of, respectively, the upper and lower layers.

In a compartment,

$$\rho_i = \frac{m_i}{V_i} \tag{6.1}$$

$$E_i = c_v(T)m_iT_i \tag{6.2}$$

$$p = \rho_i RT_i \tag{6.3}$$

$$V = V_U + V_L \tag{6.4}$$

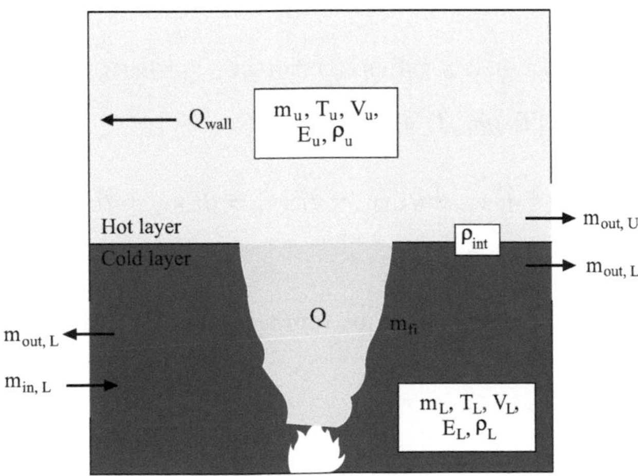

Figure 6.1 Schematic diagram of two-zone model.

where

$C_v(T)$ is the specific heat of the gas in the compartment,
R is the universal gas constant,
i takes U for upper layer and L for lower layer.

Inside each zone, constant volume and constant pressure are presumed; therefore, the specific heat of the gas at the universal gas constant R and the ratio of specific heat are related by

$$R = c_p(T_i) - c_i(T_i) \tag{6.5}$$

$$\gamma(T_i) = \frac{c_p(T_i)}{c_v(T_i)} \tag{6.6}$$

Therefore, variation of the specific heat of the gas is

$$c_p(T) = 0.187T + 952\left[\frac{\text{J}}{\text{kg K}}\right] \tag{6.7}$$

The mass balance equations in each zone are made of the mass exchanges of one zone, with the fire, with the other zone, and with the external world through the different vent types:

$$\dot{m}_U = \dot{m}_{U,VV_{\text{out}}} + \dot{m}_{U,HV_{\text{in}}} + \dot{m}_{U,HV_{\text{out}}} + \dot{m}_{U,FV_{\text{in}}} + \dot{m}_{U,FV_{\text{out}}} + \dot{m}_e + \dot{m}_{fi} \tag{6.8}$$

$$\dot{m}_L = \dot{m}_{U,VV_{\text{in}}} + \dot{m}_{L,VV_{\text{in}}} + \dot{m}_{L,VV_{\text{out}}} + \dot{m}_{L,HV_{\text{in}}} + \dot{m}_{L,HV_{\text{out}}} + \dot{m}_{L,FV_{\text{in}}} + \dot{m}_{L,FV_{\text{out}}} - \dot{m}_e \tag{6.9}$$

The energy balance equations in each zone are made of the energy exchanges of one zone, with the fire, with the other zone, with the surrounding partitions, and with the external world trough vents:

$$\dot{q}_U = \dot{q}_{U,\text{rad}} + \dot{q}_{U,\text{wall}} + \dot{q}_{U,VV_{\text{out}}} + \dot{q}_{U,HV_{\text{in}}} + + \dot{q}_{U,HV_{\text{out}}} + + \dot{q}_{U,FV_{\text{in}}} + \dot{q}_{U,FV_{\text{out}}} + c_p(T_L)\dot{m}_{\text{ent}}T_L + 0.7\text{RHR} \tag{6.10}$$

$$\dot{q}_L = \dot{q}_{L,\text{rad}} + \dot{q}_{L,\text{wall}} + \dot{q}_{U,VV_{\text{in}}} + \dot{q}_{L,VV_{\text{in}}} + \dot{q}_{L,VV_{\text{out}}} + \dot{q}_{L,HV_{\text{in}}} + \dot{q}_{L,HV_{\text{out}}} + \dot{q}_{L,FV_{\text{in}}} + \dot{q}_{L,FV_{\text{out}}} - \dot{q}_{\text{ent}} \tag{6.11}$$

Based on the above equations, zone temperatures T_U and T_L can be solved:

$$\Delta p = \frac{(\gamma - 1)\dot{q}}{V} \tag{6.12}$$

$$\dot{T}_U = \frac{1}{c_p(T_U)\rho_U V_U}\left(\dot{q}_U - c_p(T_U)\dot{m}_U V_U + V_U \Delta\dot{p}\right) \tag{6.13}$$

$$\dot{T}_L = \frac{1}{c_p(T_L)\rho_L V_L}\left(\dot{q}_L - c_p(T_L)\dot{m}_L V_L + V_U \Delta\dot{p}\right) \tag{6.14}$$

$$\dot{Z}_S = \frac{1}{\gamma(T_L) P A_f}\left(\left(\gamma(T_L) - 1\right)\dot{q} + V_U \Delta\dot{p}\right) \tag{6.15}$$

where
Z_S is the altitude of separation of zones,
$\Delta\dot{p}$ is the difference of pressure from the initial time.

6.2.2.1.2 One-zone model

In one-zone model, atmosphere temperature, density, and internal energy and pressure are assumed to be uniform in one compartment. One-zone models can be used for post-flashover fires that consider uniform distribution of temperature inside the compartment (see Figure 6.2).

In Figure 6.2,
m_g is the mass of the gas of whole compartment,
T_g is the temperature of the gas of whole compartment,
V is the volume of the compartment (constant),
E_g is the internal energy in the compartment,
p is the pressure in the compartment,
r_g is the gas density in the compartment.
In a compartment,

$$\rho_g = \frac{m_g}{V} \tag{6.16}$$

$$E_g = c_V(T_g) m_g T_g \tag{6.17}$$

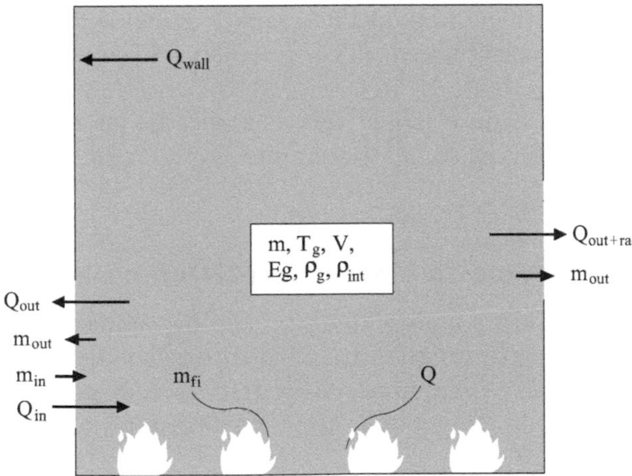

Figure 6.2 One-zone fire model.

$$p = \rho_g R T_g \tag{6.18}$$

$$V = \text{cst} \tag{6.19}$$

The mass balance is expressed as follows:

$$\dot{m}_g = \dot{m}_{\text{in}} + \dot{m}_{\text{out}} + \dot{m}_{fi} \tag{6.20}$$

And the energy balance is expressed as follows:

$$\dot{q}_U = \dot{q}_{\text{rad}} + \dot{q}_{\text{wall}} + c_p\left(T_g\right)\dot{m}_{\text{out}}T_g + c_p\left(T_{\text{out}}\right)\dot{m}_{\text{in}}T_{\text{out}} + \text{RHR} \tag{6.21}$$

Based on the above equations, zones temperatures T_U and T_L can be solved:

$$\Delta p = \frac{(\gamma - 1)\dot{q}}{V} \tag{6.22}$$

$$\dot{T}_g = \frac{1}{c_p\left(T_g\right)\rho_g V}\left(\dot{q} - c_p\left(T_g\right)\dot{m}_g T_g + V\Delta\dot{p}\right) \tag{6.23}$$

where
$\Delta\dot{p}$ is the difference of pressure from the initial time.

6.2.2.2 Limitations of zone modeling

The major drawback of zone models is the a priori knowledge of the structure of the flow. This means that the validity of assumptions involved in zone modeling should be confirmed in each particular case. This means that zone model can never be decoupled from supporting experimental studies (Novozhilov, 2001).

Zone model is difficult to model in a rapidly growing fire, due to the fact that there is no sufficient time for flow restructuring so that different zones can develop.

The zone model cannot model very complicated geometry, as it is not appropriate to model the compartment into several regular-shape compartments any more.

6.2.2.3 Computational fluid dynamics (CFD) fire modeling

CFD models are most sophisticated and can be numerically very expensive. Filed modeling techniques are used to predict the spread of smoke and heat from fires. Computationally, it considers not only the complete dependency of temperature-time relationships but also time and space dependency.

6.2.2.3.1 *Basic theories*

CFD models are the most advanced and sophisticated fire modeling technique to predict fire growth and compartment temperatures. CFD models solve the numerical solutions of the Navier-Stokes equations. The solutions of partial differential equations for conservation of mass, momentum, and energy are approximated as finite differences over a number of control volumes describing the fluid flow and heat transfer phenomena associated with fires. They can predict smoke and heat movement in buildings in the most accurate manner. CFD models allow the simulation of a large range of boundary conditions, and the ability to model much more detailed geometries than can be achieved using zone models. CFD models are capable of modeling pre-flashover and localized fires in complex geometries with smoke movement in multi-compartments. Therefore, they are commonly used for the most complicated fire engineering projects.

6.2.2.3.2 *Limitations of CFD*

The drawbacks of CFD models include increased complexity for user, greater computational requirements for hardware, and computationally time consuming.

6.2.2.3.3 *Analysis software*

Commercial CFD software packages are ANSYS-CFX, ANSYS-Fluent, Star-CD, Flow3D, CFDRC, and AVL Fire.

Open-source computational packages are FDS, OpenFoam, SmartFire, and Sophie.

6.3 DETERMINING MEMBER TEMPERATURE

In a structural fire design, after knowing the compartment (also called atmosphere or gas) temperature, the member temperature in fire condition needs to be determined. This is primarily through heat transfer.

6.3.1 Simplified temperature increase models from Eurocode

As introduced in Chapter 3, for internal members in fire compartments, a method for the calculation of the member temperature is given in Eurocode 3 (EN 1993-1-2, 2005) and Eurocode 4 (EN 1994-1-2, 2005) for both protected and unprotected structural members.

For external members, the radiative heat flux component should be calculated as the sum of the contributions of the fire compartment and of the flames emerging from the openings; a method for the calculation of the heating conditions is given in BS EN 1991-1-2: Eurocode 1 annex B.

6.3.2 Heat transfer using finite element method

6.3.2.1 Theoretical principles

In finite element method, when performing heat transfer analysis, the investigated body is first divided into elements. Temperature is the function of each node in an element (such as shell or beam elements). By expressing the temperature field in terms of assumed interpolation functions of each node within each element, the temperature of the element can be obtained. The interpolation functions are defined in terms of the values of the temperature field at specified nodes. The nodal temperature field and the interpolation functions for the elements completely define the behavior of the temperature field within the elements.

The finite element heat transfer analysis is to solve the nodal temperature. The matrix equations of the individual elements are determined from the governing equation by using the weighted residual approach. The individual element matrix equations are then combined to form the global matrix equations for the complete system. Once the boundary conditions have been imposed, the global matrix equations can be solved numerically. Once the nodal temperature for each element is found, the interpolation functions is used to determine the temperature field throughout the whole assemblage of elements. The global finite element matrix equation is obtained by the assembly of element matrix equations, which can be expressed as follows (Purkiss, 2007):

$$C\frac{\partial\theta(t)}{\partial t}+K_c\theta(t)=R_q+R_Q \tag{6.24}$$

where

$$C=\sum_{e=1}^{M}C_c=\text{global capacitance matrix}$$

$$K_c=\sum_{e=1}^{M}K_{ce}=\text{global conductance matrix}$$

$$R_q=\sum_{e=1}^{M}R_{qe}=\text{global nodal vector of heat flow}$$

$$R_Q=\sum_{e=1}^{M}R_{Qe}=\text{global nodal vector of internal heat source}$$

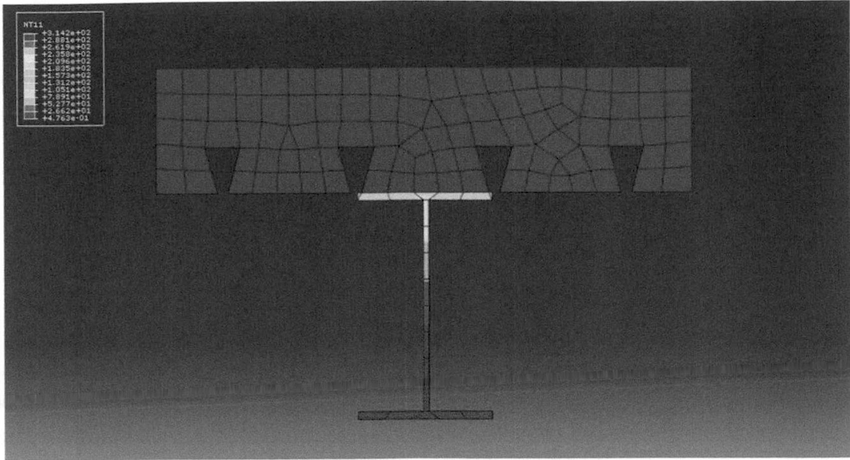

Figure 6.3 Temperature distribution of a composite beam after heat transfer analysis usingAbaqus®.(Abaqus®screenshotreprintedwithpermissionfromDassaultSystèmes.)

6.3.2.2 Analysis software and modeling example

Most of the commercial software have the capacity to perform heat transfer analysis. Software packages such as Abaqus, ANSYS, and LS-DYNA can all perform heat transfer analysis.

In finite element analysis, either standard fire temperature curve or parametric fire temperature curve is used to represent the atmosphere fire development inside a compartment. As shown in Figure 6.3, a heat transfer analysis is performed for a composite beam in Abaqus®. A parametric fire temperature was used for the analysis. In Abaqus, a special type of heat transfer element should be chosen to enable the heat transfer analysis. As demonstrated in Figure 6.3, it presents the temperature profile of a composite beam heated up from bottom of the steel section. For a detailed heat transfer procedure in Abaqus, refer to Fu (2015).

Heat transfer analysis can only determine the temperature distribution of the structural members, but not the structural response, such as mechanical stress and deformation. To be able to obtain the structural response, thermal-mechanical coupled analysis or partial thermal-mechanical coupled analysis is needed, which will be introduced in detail in Section 6.4.

6.4 DETERMINING STRUCTURAL RESPONSE OF STRUCTURAL MEMBERS IN FIRE

After knowing the temperature distribution, the structural response of the structural members needs to be solved. There are several methods to solve the structural response in fire which will be introduced here.

6.4.1 Multi-physics fire analysis (thermal mechanical coupled analysis)

Most software can perform independent thermal and stress analysis. However, in the event of fire, the stress and deformation change with the changes of the temperature. Conversely, the mechanical change will also cause change to the temperature field. Therefore, a fully coupled thermal mechanical analysis considering the mutual effect will provide the most accurate modeling result.

Multi-physics fire analysis is also called coupled stress-thermal analysis, real heat transfer analysis, or thermal-mechanical coupled analysis. Thermo-mechanical problems are usually solved by employing 3D finite elements available in commercial finite element software. Most of these codes are based on partially coupled thermoelastic formulations, which means that they can provide the mechanical stress and deformation due to a thermal loading but do not allow the computation of the temperature change caused by a mechanical loading.

A fully coupled thermal-mechanical analysis should allow the mutual computational exchange of both thermal loading to mechanical deformation and temperature change caused by mechanical loading. Most commercial software has the capacity of fully coupled heat transfer analysis, such as ANSYS, Abaqus®, and ADINA.

6.4.1.1 Abaqus®

The latest version of Abaqus® CAE/Standard can perform fully coupled thermal-mechanical analysis through defining coupled temp-displacement analysis step. As shown in Figure 6.4, this step can be defined in step module. In the analysis, coupled temp-displacement elements should be used to enable the analysis (as shown in Figure 6.5). This is different to heat transfer analysis just been introduced, where heat transfer elements should be chosen.

6.4.1.2 ADINA

ADINA TMC is primarily used for coupled thermo-mechanical problems. The program can handle fully coupled problems where the thermal solution affects the structural solution and the structural solution also affects the thermal solution.

1. Mechanical to thermal coupling:
   Internal heat generation due to plastic deformations or viscous effects
   Heat transfer between contacting bodies
   Heat generation due to friction

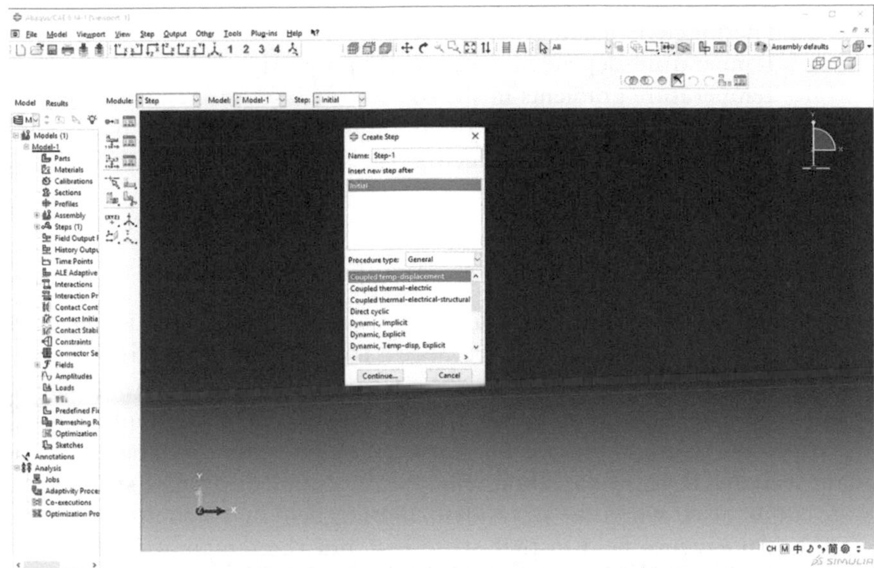

Figure 6.4 Define coupled temp-displacement analysis step in Abaqus®.(Abaqus® screenshot reprinted with permission from Dassault Systèmes.)

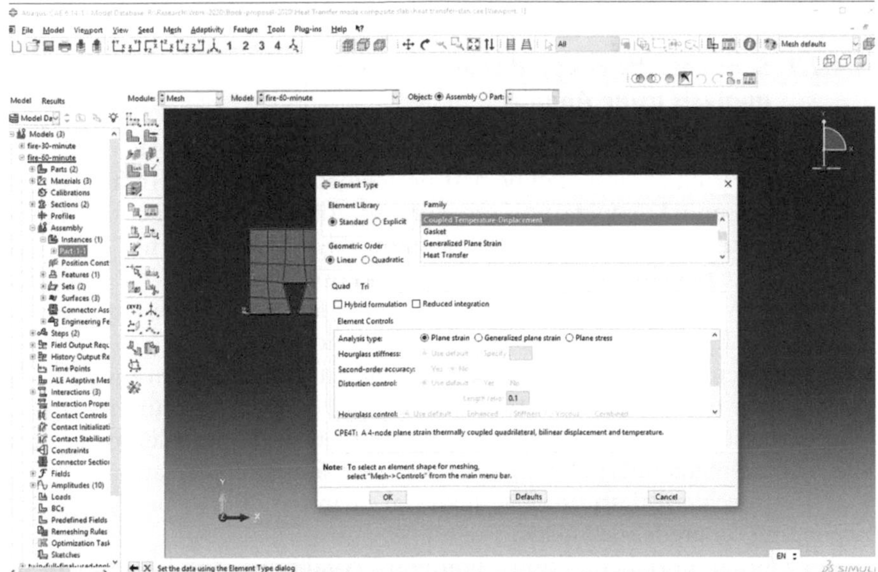

Figure 6.5 Define coupled temp-displacement element type in Abaqus®. (Abaqus® screenshot reprinted with permission from Dassault Systèmes.)

2. Thermal to mechanical coupling:
   Thermal expansion
   Temperature-dependent mechanical properties
   Temperature gradients in shells.

ADINA TMC can handle one-way coupling where only one of the two physics fields affects the other which is the so-called sequential modeling that will be introduced in the next section.

6.4.2 Sequentially coupled thermal-stress analysis

Fully coupled thermal-mechanical analysis is very expensive in terms of computational cost. It may not be necessary in most of the fire analysis for tall buildings as the change of the temperature due to the mechanical change may not significantly affect the structural fire design in a tall building. Therefore, a more computational effective method, partial thermal-mechanical or the so-called sequentially coupled thermal-stress analysis, can be used when the stress/deformation field in a structure depends on the temperature field in that structure, but the temperature field can be found without the knowledge of the stress/deformation response. The analysis is usually performed by first conducting an uncoupled heat transfer analysis and then a stress/deformation analysis. In this analysis, the temperature field does not depend on the stress field. Most commercial software have the capacity of sequentially coupled heat transfer analysis, such as ANSYS and Abaqus®.

6.4.2.1 Sequentially coupled thermal-stress analysis using Abaqus®

Sequentially coupled thermal-stress analysis is available in **Abaqus®**. The heat transferring analysis can be first performed in **Abaqus®**. In heat transfer analysis, nodal temperatures are stored in **Abaqus®** as a function of time in the heat transfer results (.fil) file or output database (.odb) file. The temperatures are read into the stress analysis as a predefined field. The temperature varies with position and is usually time dependent. The temperature field is not changed by the stress analysis solution.

The element temperature is then interpolated to the calculation of nodal temperature obtained in heat transfer analysis. The temperature interpolation in the stress elements is usually approximate and one order lower than the displacement interpolation to obtain a compatible variation of thermal and mechanical strain. The material degradation can be determined depending on predefined fields.

Appropriate initial conditions, boundary conditions, and loading are set in both the heat transfer and subsequent stress analysis. Thermal strain will arise in the stress analysis due to the thermal expansion included in the material property definition. Any of the heat transfer elements in ABAQUS (Standard) can be used in the thermal analysis. In the stress analysis, the corresponding continuum or structural elements must be chosen. One should *choose the element type compatible with the heat transferelement type used.*

Below is an inp code example of sequential analysis in ABAQUS.

```
Heat transfer analysis
"*HEADING
...
*ELEMENT, TYPE=SR4 (heat transfer element type should be
chosen)
*STEP
*HEAT TRANSFER(perform heat transfer analysis)
...
**
*NODE FILE, NSET=NALL
 NT
*OUTPUT, FIELD
*NODE OUTPUT, NSET=NALL
 NT (Write all nodal temperatures to output database file,
heat.fil/heat.odb)
*END STEP
Static structural analysis
*HEADING
...
*ELEMENT, TYPE=SR4
(Choose the element type compatible with the heat transfer
element type used)
...
*STEP
*STATIC(perform heat transfer analysis under the condition
of fire)
*TEMPERATURE, FILE=heat
Read in all nodal temperatures from the results or output
database file, heat.fil/heat.odb
...
*END STEP"
```

6.4.2.2 Sequentially coupled thermal-stress analysis using ANSYS

ANSYS Workbench Platform can also perform sequentially coupled thermal-stress analysis. As shown in Figure 6.6, a steady-state thermal analysis is first set up, which is linked to another module called static structural analysis module.

To clearly demonstrate the whole modeling process, a real construction project, the long span roof of Kings Cross station (see Figure 6.7) is

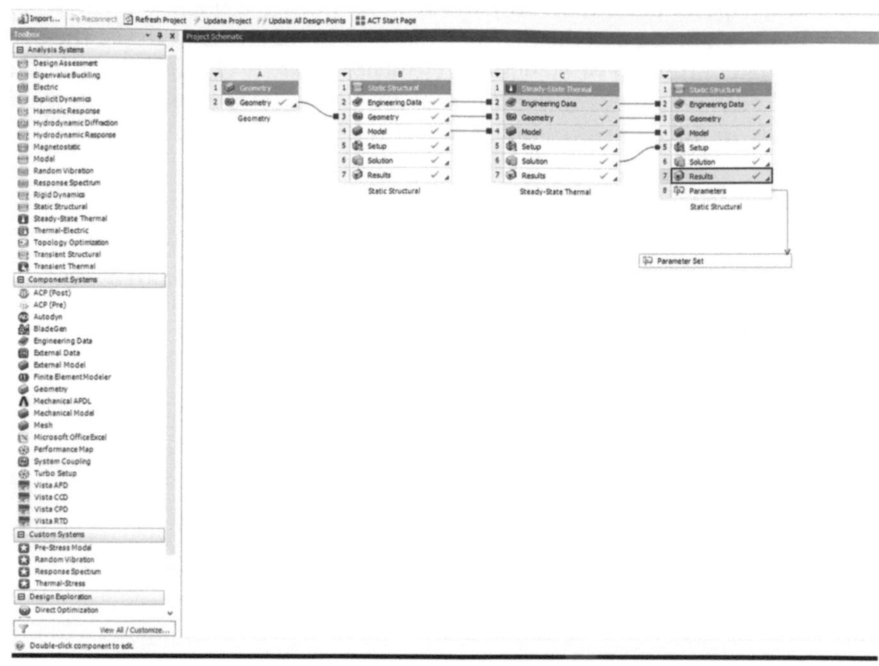

Figure 6.6 Thermal-structural analysis using ANSYS. (Screenshot reprinted with permission from ANSYS, Inc.)

Figure 6.7 Kings Cross station roof (U.K.). (Photo taken by the author.)

simulated in ANSYS. The following figure shows the steps of the thermal-structural analysis using ANSYS. To simplify the analysis, a standard fire temperature is adopted as atmosphere temperature in the analysis.

As shown in Figure 6.8, the steady-state thermal analysis is set up through defining the convection, radiation, and conduction (see the left column in Figure 6.6). A standard fire temperature curve from Eurocode has been adopted as the atmosphere temperature.

As shown in Figure 6.9, after steady-state thermal analysis, the result of a temperature field can be obtained.

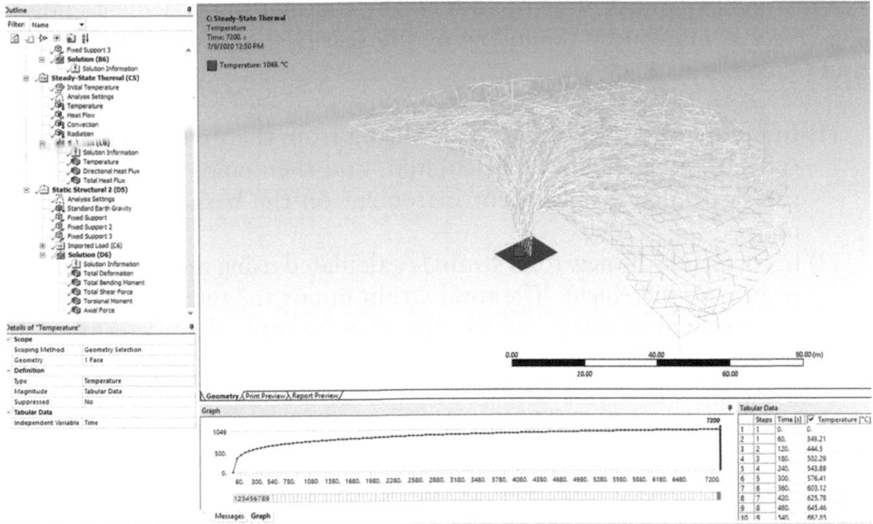

Figure 6.8 Steady-state thermal analysis set-up. (Screenshot reprinted with permission from ANSYS, Inc.)

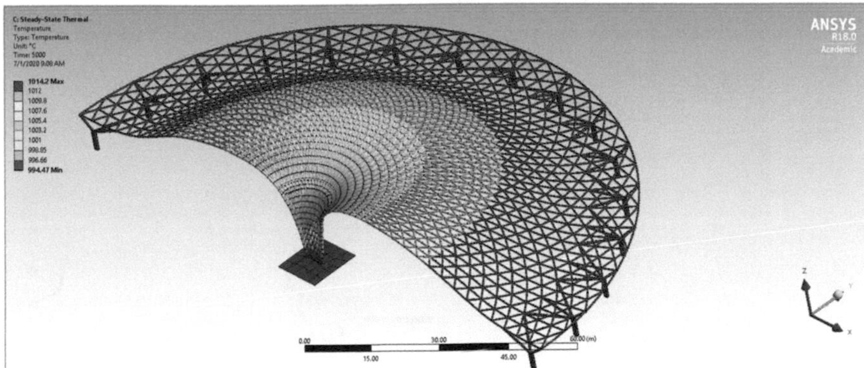

Figure 6.9 Temperature field result from steady thermal analysis in ANSYS. (Screenshot reprinted with permission from ANSYS, Inc.)

The results of thermal analysis are then transferred to the structural model with the capability of ANSYS to use the same model for the two analyses. Figure 6.10 shows the thermal-structural analysis results of a dome using ANSYS. It shows the deformation of the whole roof after combined fire and gravity load.

6.4.2.3 Partial thermal-mechanical analyses in OpenSees

A partial thermal-mechanical analysis was available in OpenSees for fire. It can undertake the mechanical component of a decoupled analysis in the fire. As explained by Jiang (2012), the thermal finite element code in OpenSees is marked by the following phases: (1) predictor, (2) corrector, and (3) convergence check.

(1) In predictor, unbalance force resulting from thermal load and material softening should be calculated, and then employ it to compute the associated displacement increment on the basis of the stiffness matrix.
(2) In corrector, the new total strain is calculated using the predictor's displacement increment. The total strain minus the thermal strain gives the mechanical strain, which can then be utilized to determine the stress state. The integration of stresses across the cross section produces the out-of-balance force.
(3) In convergence check, this involves assessing the out-of-balance forces, after which the attainment of convergence occurs upon equilibrium in the structure. This stage is consistent with the standard processes involved in the original OpenSees.

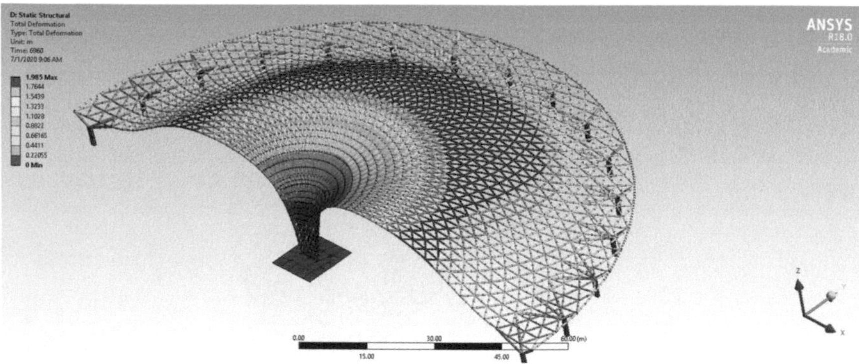

Figure 6.10 Result of deflection after structural analysis using ANSYS. (Screenshot reprinted with permission from ANSYS, Inc.)

6.4.2.4 *Codified thermal-mechanical coupled analysis*

6.4.2.4.1 Finite element package

Fu (2016a, b) developed a new codified finite element analysis method. It can perform 3D finite element structural fire analysis based on Eurocode equations for temperature increase of steel members and heat transfer analysis for concrete member.

In the first step, a 3D structural finite element model is built.

In the second step, gas temperatures in the form of a fire temperature time curve according to Eurocode 1 part 1–2 (EN 1991-1-2, 2002) is used to determine the temperature increase of the protected or unprotected structural members based on equations from design guidance such as Eurocode 3, Part 1–2 (EN 1993-1-2,2005) and Eurocode 4, Part 1–2 (EN 1994-1-2,2005).

In the third step, the increase of nodal temperature of the slab (coming from a heat transfer analysis) and the protected or unprotected beams and columns (calculated using the formula from Eurocode 3 (EN 1993-1-2, 2005)) is applied to the correspondent structural members.

```
Fourth step—partially coupled thermal-mechanical fire
analysis
```

The partial coupled thermal-mechanical fire analysis will be defined as follows:

```
****Step2*********
*step,unsymm=yes,nlgeom=yes,inc=5000
*static
60,600,1e-30,60
*temperature
slab-fire,20.1014,163.625,700.431,37.2394,22.0434 (defining
the temperature of the slabs,)
**protected beam (defining the temperature of the protected
beam and column)
probeam,48.3490677963654
procolumn,49.2077087054519
**unprotected beam (defining the temperature of the
unprotected beam and column)
beam-fire,60.5778721012576
column-fire,46.0727787944676
*end step
```

The stress and deformation due to temperature increase will be worked out.

The number of coupled thermal-mechanical fire analysis steps depends on the length of fire duration. For an instance, if the duration of the fire is 2 h, the whole fire duration can be divided into 120/10=12 steps, and the temperature within 10 min of time is presumed to be constant. More steps can be chosen for a closer representation of the real fire development.

To clearly demonstrate this process, WTC1 of Twin Tower is modeled in Abaqus®. As shown in Figure 6.11, the aircraft hit the towers at high speed impacting between the 93rd and 99th floors, causing the subsequent fire on these floors. Therefore, fire was set to levels 93–96 in Abaqus® (as shown in Figure 6.12).

The temperature of the slab is determined through the heat transfer analysis. The temperature of steel members is determined using the equations from Eurocode 3, Part 1–2 (EN 1993-1-2, 2005) and Eurocode 4, Part 1–2 (EN 1994-1-2, 2005). After the temperature is determined, thermal-mechanical analysis is performed. The results such as deformation and member force

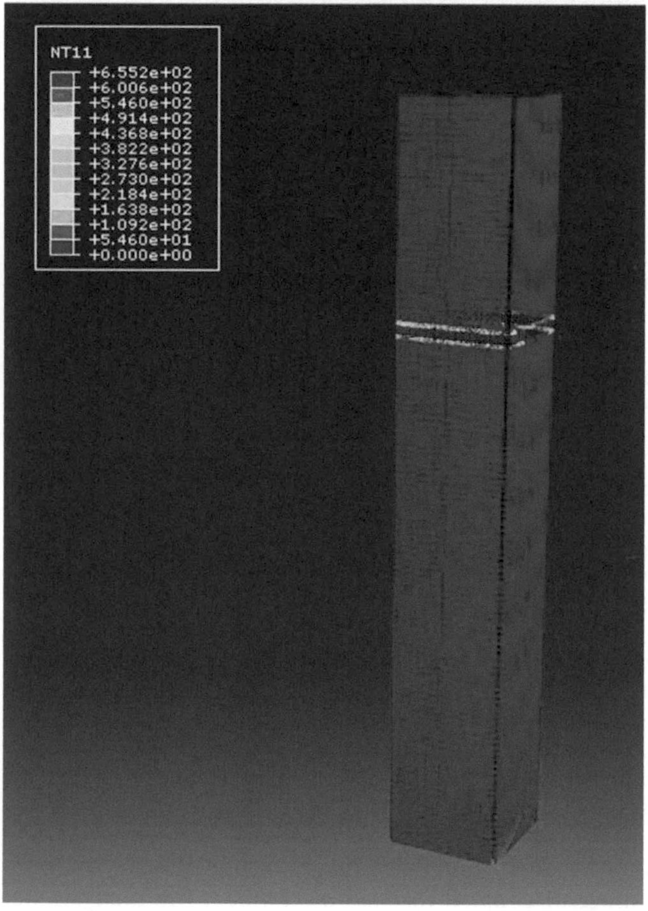

Figure 6.11 3D finite element model of WTC1. (Abaqus® screenshot reprinted with permission from Dassault Systèmes.)

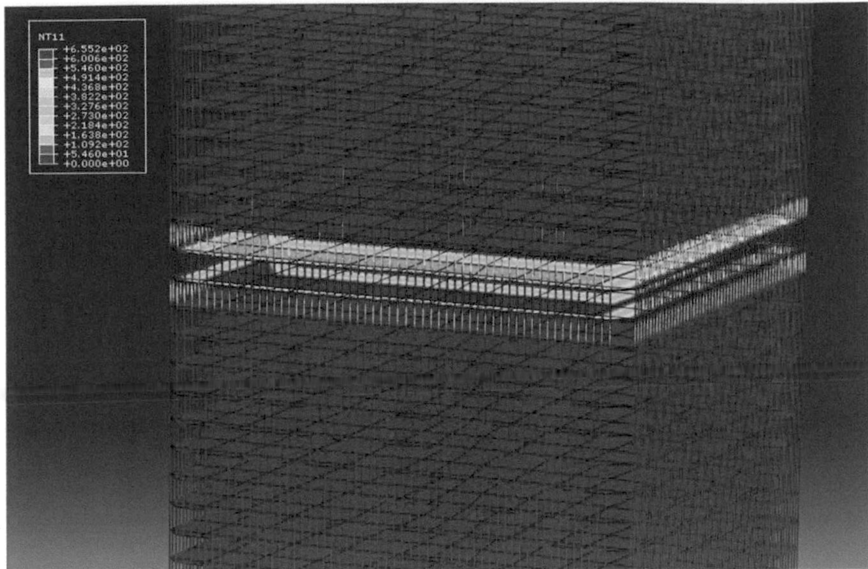

Figure 6.12 Fire was set to floors 93–96 in the model. (Abaqus® screenshot reprinted with permission from Dassault Systèmes.)

can therefore be determined. Figure 6.13 shows the vertical deformation of the whole tower in fire. It can be seen that due to the buckling or yielding of the columns, the floor above level 96 shows greater vertical deflection than the floors below level 93, showing the potential of the collapse.

6.5 PROBABILISTIC METHOD FOR FIRE SAFETY DESIGN

The probabilistic method has been increasingly used for fire safety design. However, it is based on the repeated evaluation of the structural behavior under fire loading, which is computationally expensive even for simple structural models, thereby severely hindering probabilistically.

6.5.1 Reliability-based structural fire design and analysis

Reliability-based method has been used for design codes to determine partial load factors and material safety factors. They are derived from First-Order Reliability Methods (FORM) with the intention of ensuring that structural elements or sub-frame assemblies have an appropriately low probability of failure. EN 1990:2002+A1 (2005) specifies that all design solutions should

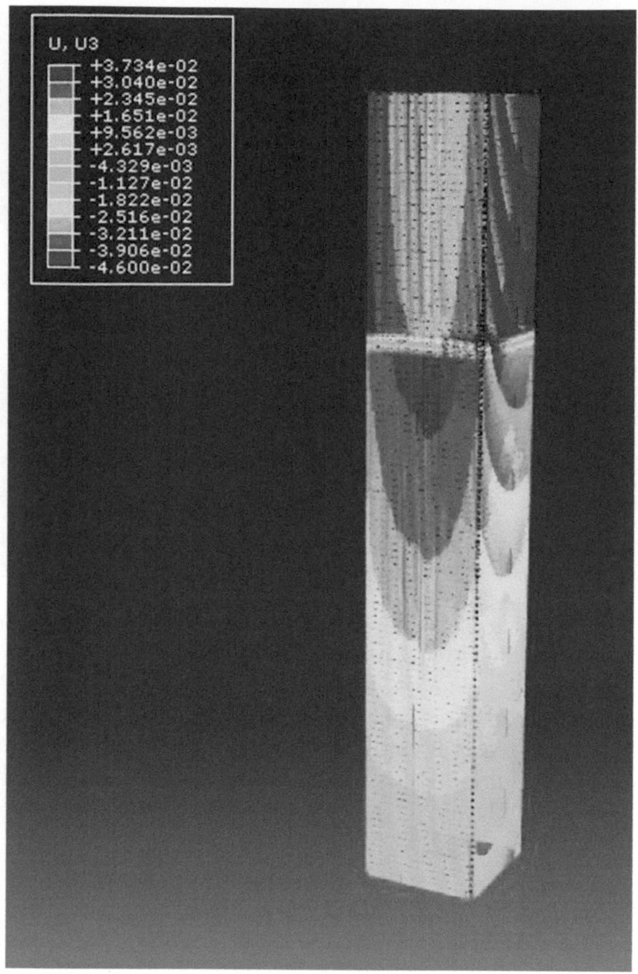

Figure 6.13 Vertical deformation distribution of WTC1 in fire. (Abaqus® screenshot reprinted with permission from Dassault Systèmes.)

achieve a reliability index (β) of 3.8 in a building's conceptual design life (50 years).

6.5.1.1 The basic reliability design principles

In structural fire, design reliability method to assess the fire resistance (capacity) of a building element or structure is sufficient to withstand the fire severity (demand). Due to the complex nature of this problem, the available capacity and demand may be modeled as random variables. The reliability of either a building system or a structural member can be measured in terms of probability. The probability of failure P_f is thus expressed as follows:

$$P_f = P(R - S \leq 0) = P(M \leq 0) \tag{6.25}$$

where
R=fire resistance (capacity),
S=fire severity (demand),
M=safety margin defined as M=R – S.
The equation M=R–S is also known as the **limit state function.**

The objective of a reliability-based design is to ensure that given the outbreak of any fire, the event (M>0) or $G(R,S)$>0. This assurance is only possible in terms of the probability $P(M>0)$. This probability therefore represents a realistic measure of the reliability of the system or a structural member in fire.

If the probability density functions $f_R(r)$ and $f_s(s)$ of R and S are available or can be approximated, and if R and S are statistically independent, the probability of failure P_f may can be expressed as follows:

$$P_f = P((R - S \leq 0) = \int_{-\infty}^{\infty} \int_{-\infty}^{n \geq r} f_R(r) f_s(S) drds \tag{6.26}$$

The reliability-based structural fire design and analysis are to work out P_f shown in Equation 6.26. The procedure for reliability-based design and analysis will be introduced in the following sections.

6.5.1.2 Reliability-based design and analysis procedure

6.5.1.2.1 Determination of limit state function

The limit state function M=R–S needs to be determined first. This is based on the type of design problems (whether to design the flexural capacity of a beam under fire or the overall stability of a building in fire).

6.5.1.2.2 Monte Carlo simulation method

The Monte Carlo technique is applicable for either stochastic or probabilistic problem. The process is computational and involves selecting input values at random for use in engineering calculations. Monte Carlo simulation is a choice of probability distributions for the random inputs. It uses the randomness to solve problems that might be deterministic in principle. Monte Carlo methods can be used to sample using a known probability distribution.

It first needs to select a probability distribution for each individual variable. It is also essential to determine the dependencies between simulation inputs. Ideally, input data to a simulation should reflect what is known about dependence among the real quantities being modeled. Probability distributions for each variable need to be created from statistical data or information taken from real-life observations or experimental data.

In the fire safety design, there are some key parameters that will affect the design values. For example, when determining atmosphere temperature, opening factor and fire load density are the two key parameters to affect its value, and they are mutually independent to each other. The probability distribution or the range of these parameters is readily known from design guidelines such as Eurocode and other research as shown in Table 6.1. Therefore, using the available distributions and key statistic index such as mean and standard deviation obtained from Eurocodes design practice (see Table 6.1), the random value of opening factor and fire load density can be generated. Subsequently, the corresponding atmosphere temperature can be calculated based on design formula. The random variables used are determined from large-scale data analysis and tests.

6.5.1.2.3 Determining statistical parameters of the variables

The statistical parameters of design are given by Eurocode 1 (2002) and Eurocode 3 (2005). Using these statistic parameters or the range of the design parameters, the values of the variable can be generated using Monte Carlo simulation. Table 6.1 shows examples of statistical parameters of the variables used in the Monte Carlo simulation.

Table 6.1 Examples of statistical parameters of the variables used in the limit function

Variable	*Distribution*	*Units*	*Mean*	*Standard deviation*	*Range*	*Source*
Opening factor	Normal	N/A	N/A	N/A	0.02–0.2	Eurocode 1; Part 1.2 (2002)
Fire load density	Gumbel	MJ/m^2	420	126		Eurocode 3 (2005)
Imposed load	Extreme type I	KN/m^2			1–5	Eurocode 1; Part 1.2 (2002)
Yield strength of steel	Log-normal	MPa	280	28	275–355	Eurocode 1; Part 1.2 (2005)
Partial safety factors	Normal	-	-	-	1.5–2	Eurocode 1; Part 1.2 (2002)

6.5.1.3 *Case study for reliability analysis for individual members*

Cai and Fu et al. (2020) developed a new approach for post-fire reliability analysis of concrete beams retrofitted with CFRP sheet in bending using the Monte Carlo method. It is introduced here.

6.5.1.3.1 *Limit state function*

The limit state function of a beam in bending in ambient temperature can be written as follows:

$$Z = R - S = g(X_1, X_2, \ldots, X_n) \tag{6.27}$$

where $g(X)$ is the failure function, X_1, X_2, ..., X_n are n mutually independent random variables, R is the resistance of the structure, and S is the action effect of the structure. Values of Z greater than 0, less than 0, or equal to 0 indicate that the structure is under a reliable status, a failure status, or a limit status, respectively.

The flexural capacity of RC beams at ambient temperature was (GB, 2010) as follows:

$$M_C = \alpha_1 f_c bx(h_0 - 0.5x) + f_y' A_s'(h_0 - a_s') \tag{6.28}$$

where M_C is the flexural capacity of RC beams at normal temperature. With a random variable γ_m that represents the uncertainty coefficient of the resistance calculation, the limit state function of RC beams at normal temperature is as follows:

$$Z = R - S = \gamma_m M_C - (M_{Gm} + M_{Qm}) \tag{6.29}$$

where M_{Gm} is the mean value of M_G and M_{Qm} is the mean of M_Q.

The limit state function of RC beams after fire is as follows:

$$Z = R - S = \gamma_m M_{CT} - (M_{Gm} + M_{Qm}) \tag{6.30}$$

Since the flexural capacity of RC beams was deteriorated after fire exposure, CFRPs can be used to reinforce the bottom of post-fire RC beams. If the bonding between CFRP and concrete is assumed to be perfect (GB, 2013), the flexural capacity of post-fire CFRP-reinforced RC beams can be computed as follows:

$$M_D = \alpha_1 \bar{\varphi}_{CT} f_c bx(h - 0.5x) + \varphi_{yT}' f_y' A_s'(h - a_s') - \varphi_{yT} f_y A_s (h - h_0) \tag{6.31}$$

$$x = (\varphi_{yT} f_y A_s + \psi_f f_{fu,s} A_{fe} - \varphi_{yT}' f_y' A_s') / (\alpha_1 \bar{\varphi}_{CT} f_c) \tag{6.32}$$

where

M_D is the flexural capacity of post-fire RC beams strengthened with CFRPs,
$f_{fu,s}$ is the mean tensile strength of CFRPs,
A_{fe} is the valid sectional area of CFRPs,
ψ_f is the strength use coefficient of CFRPs.

The limit state function of post-fire RC beams strengthened with CFRPs was determined as follows:

$$Z = R - S = \gamma_m M_D - \left(M_{Gm} + M_{Qm}\right) \tag{6.33}$$

6.5.1.3.2 Statistical parameters of the variables

The statistical parameters are selected based on the Chinese Code (GB, 2010), EN 1992-1-2(2004)), and the research of Cai (2016) and Coile et al. (2014) (Table 6.2).

6.5.1.3.3 Monte Carlo simulation

In the Monte Carlo simulation, the random variables for the limit state function were repeatedly simulated using the program coded in MATLAB, and the reliability can be calculated. The specific procedures are listed as follows:

- The random variables of the limit state function were integrated with their probability distributions.
- Random values were simulated repeatedly using the Monte Carlo method with the probability distribution of these random variables.
- $g(X)$ was calculated using the simulated values.
- When the number of repetitions reached the preset value, the simulations were terminated.
- Calculate the reliability based on the simulation results.

6.5.1.4 Case study for reliability analysis for a whole building

To assess the probability of a whole building failure under fire is more complicated. Van Coile et al. (2014) developed a method considering the following probability:

p_{ig} is the probability of a fire to develop,
$p_{f,u}$ is the probability of early intervention by the occupants,
$p_{f,s}$ is the probability active measures, such as sprinklers,
$p_{f,fb}$ is the probability of the fire brigade,
$P_{f,fi}$ is the probability of the structure damage,
Pf,1 is the probability of the structure fails.

Table 6.2 Statistical parameters of the variables used in the limit function

Symbol	Variable	Distribution	Units	Bias[a] (mean)	CoV[b] (std[c])	Nominal value	Source
f_c	C30 compressive strength	Lognormal	MPa	1.395	0.[illegible]5	20.1	GB (2010)
f_y	HRB335 steel yield stress	Lognormal	MPa	1.139	0.07	335	GB (2010)
S_Q	Live load	Extreme type I	kN·m	0.859	0.233	-	GB (2012)
S_G	Dead load	Normal	kN·m	1.060	0.070	-	GB (2012)
h_0	Effective depth of section	Normal	mm	1.000	0.030	565	CAI et al. (2016)
b	Beam width	Normal	mm	1.000	0.0[illegible]0	250	CAI et al. (2016)
A_f	CFRP cross-sectional area	Normal	mm²	1.00	0.02[illegible]	-	CAI et al. (2016)
f_f	CFRP tensile strength	Weibull	MPa	1.152	0.08	3,100	CAI et al. (2016)
γ_m	Total model uncertainty	Normal	-	1	0.025	-	Coile et al. (2014)
t_f	CFRP strip thickness	Lognormal	mm	1.00	0.010	0.167	CAI et al. (2016)
$\bar{\varphi}_{CT}$	T (°C) concrete compressive strength reduction factor	Beta	-	(Temperature-dependent, conforms to EN 1992-1-2)	(Temperature dependent)	-	EN 1992-1-2:2004
$\varphi_{yT}\left(\varphi'_{yT}\right)$	T (°C) reinforcement yield stress reduction factor	Beta	-	(Temperature-dependent, conforms to EN 1992-1-2)	(Temperature dependent)	-	EN 1992-1-2:2004

[a] Bias: mean value/nominal value.
[b] CoV: coefficient of variation.
[c] std: standard deviation.

This analysis process proposed by Van Coile et al. (2014) is shown in Figure 6.14. It comprises two domains: the 'event instigation' and 'response' domains.

Based on this framework, Hopkin et al. (2017) made a reliability analysis of the probability of failure of a tall building in fire. The Monte Carlo simulation is used to sample different variables, such as fire load density and temperature spread rate; therefore, different fire scenarios can be simulated and the probability of failure of the building can be assessed.

6.5.2 Fire fragility functions

Fragility function has been recently used by researchers to characterize the probabilistic vulnerability of buildings to fire. In earthquake engineering, fragility functions are widely used to assess the likelihood of structural damage due to an earthquake. A fragility function provides the probability of exceeding a damage state for a given intensity of earthquake load. Similarly, fire fragility functions can be developed to measure the probability of exceeding a damage state (e.g. column failure, excessive beam deflection) for a given intensity of fire load.

This method is quite new, and the influence of the different uncertain parameters on the functions has not been systematically studied. Uncertainties in fire, heat transfer and structural models, fire load intensity, compartment geometry and openings, the thickness and thermal conductivity of fire protection, and the material degradation, all generate significant variability in the fire fragility. The prevailing parameters in constructing fire fragility functions for steel frame buildings identified through sensitivity analyses are conducted using the Monte Carlo simulations and a

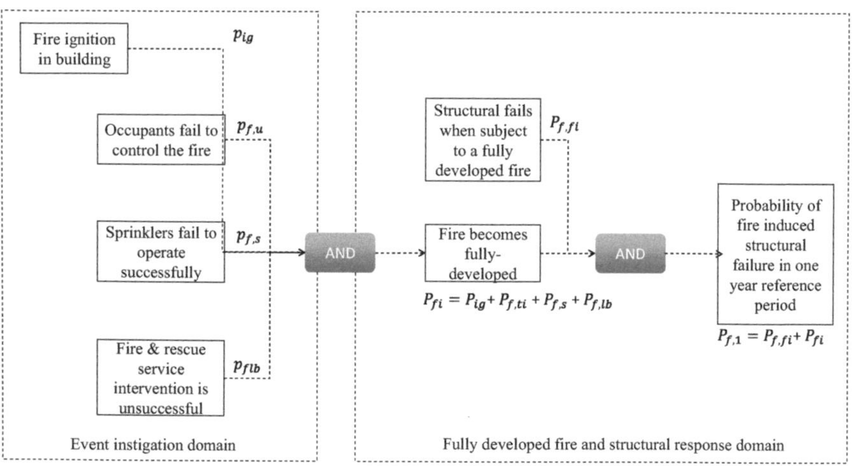

Figure 6.14 Stochastic factors leading to a fire-induced structural failure (Coile et al., 2014).

variance-based method. One important parameter in defining a fragility function is intensity measure for fire load. Gernay et al. (2019) used average fire load (in MJ/m^2 of floor area) as the intensity measure.

6.5.2.3 Compartment-level fragility function

Khorasan et al. (2016) and, Gernay et al. (2019) developed the below equation:

$$P_{F|Hfi} = \int_{0}^{\infty} \left[1 - F_{D|Hfi}(\alpha)\right] f_c(\alpha) d\alpha \tag{6.34}$$

where

$P_{F|Hfi}$ is the probability of reaching a damage state condition to the occurrence of a fire H_{fi},

$F_{D|Hfi}$ (.) is the cumulative distribution function of the demand relative to the fire H_{fi},

$f_C(\cdot)$ is the probabilistic distribution function of capacity.

$P_{F|Hfi}$ is obtained through repeated structural fire analysis under fire load densities (q values) in the same compartment. The analysis needs to be repeated sufficient times to be able to work out $P_{F|Hfi}$. Based on $P_{F|Hfi}$, the fragility function is built by fitting a function to the obtained points, assuming a lognormal distribution:

$$F(q) = \varnothing \left[\frac{\ln\left(\frac{q}{c}\right)}{\zeta} \right] \tag{6.35}$$

where

q is the fire load (MJ/m^2),

$\Phi[\bullet]$ is the standardized lognormal distribution function.

There are two parameters c and ζ:

c is the mean of lognormal distribution,

ζ is the standard deviation of the lognormal distribution.

c and ζ are determined by the best fit to the data points from structural fire analysis.

6.5.2.4 Building-level fragility function

Fire fragility functions should first be developed for each compartment under different fire scenarios and then combined to derive a fire fragility function for the entire building. The combined fragility function is also a lognormal function, the same as in Equation 6.35. Khorasan et al. (2016) and Gernay et al. (2019) developed the following equations to work out the two lognormal parameters:

$$q_c = \prod_{i=1}^{n} c_i^{p_i} \tag{6.36}$$

$$\zeta_c^2 = P^T Z + A^T Q A \tag{6.37}$$

$$Q = \begin{bmatrix} p_1(1-p_1) & \cdots & -p_1 p_n \\ \vdots & \ddots & \vdots \\ -p_n p_1 & \cdots & p_n(1-p_n) \end{bmatrix} \tag{6.38}$$

where

q_c is the mean of combined lognormal distribution,
ζ_c is the standard deviation of the combined lognormal distribution,
Other parameters are as follows:
n is the number of 'nominally identical but statistically different' fragility curves,
c_i is the median associated with each individual fragility curve,
p_i is the conditional probability for a fire in compartment,
i, a fire occurs in the building,
P is the vector of the probabilities p_i,
Z is the vector of the variances,
ζ_i^2 is associated with each individual fragility function,
A is the vector of the expected values ($\ln c_i$), and
Q is the matrix given by $Q = \begin{bmatrix} p_1(1-p_1) & \cdots & -p_1 p_n \\ \vdots & \ddots & \vdots \\ -p_n p_1 & \cdots & p_n(1-p_n) \end{bmatrix}$

Based on the above approaches, fire fragility curve can be derived. Figure 6.15 shows typical fire fragility curves for different types of buildings.

6.5.3 Other probabilistic approaches in fire safety design

Fu (2020) used the Monte Carlo simulation to simulate a probability distribution for different variables for the random inputs for fire-induced tall building collapse using machine learning. This will be introduced in detail in Chapter 9.

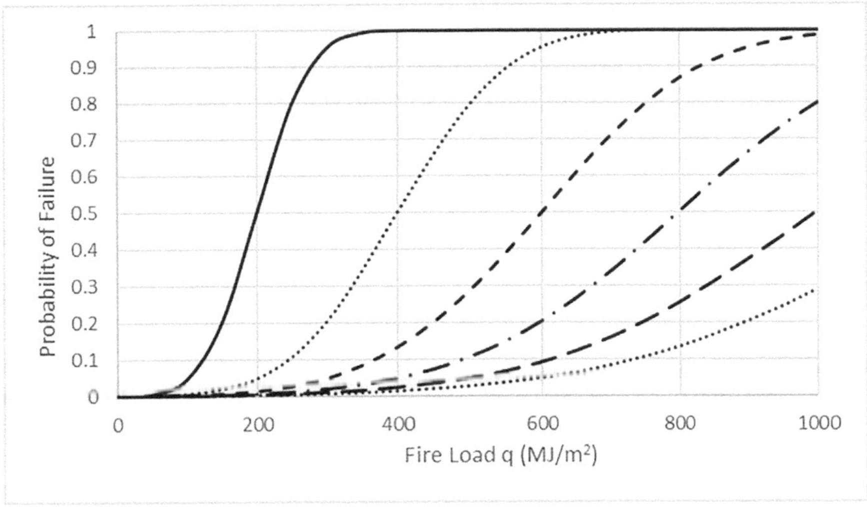

Figure 6.15 Fire fragility curves for different types of buildings.

6.6 MAJOR FIRE ANALYSIS SOFTWARE

There are several fire analysis software packages available on the market for engineers to choose. Research-focused software developed by certain academic research groups, including VULCAN, ADAPTIC, and SAFIR, focus on highly specific modeling issues. They are not widely used by engineers, due to their availability. Conversely, many commercial software packages, including DIANA, ANSYS, and ABAQUS, are widely used by engineers, but the source code is proprietary and expensive to purchase.

Some open-source codes such as OpenSees can be freely accessed, and users can develop the software package based on their needs. It also allowed for the modification of stiffness matrices and load vectors, as well as develop new element. This is usually difficult in commercial software. In terms of fire, it has a library of temperature-dependent materials and thermal elements for both shells and beams. It is also able to perform thermal-mechanical analysis. ABAQUS and ANSYS have been introduced in the preceding sections of this chapter; therefore, in this section, other major fire analysis software will be introduced.

6.6.1 Ozone software

Ozone software is developed by the University of Liege, Belgium. It was developed primary based on the theory of zone models. It includes a two-zone model and a one-zone model with a possible switch from two to one zone if some criteria are encountered. Therefore, it can model both localized and fully developed fires. It can also solve the member temperatures, such as wall model which is made by the implicit finite element method. Different combustion models have been developed to cover different situations of use of the code. A Graphic User Interface has been developed to define the input data.

6.6.2 CFAST

CFAST is a free and open-source software provided by the National Institute of Standards and Technology (NIST) of the United States Department of Commerce. CFAST is another two-zone fire model capable of predicting the environment in a multi-compartment structure subjected to a fire. It calculates the time-evolving distribution of smoke and gaseous combustion products as well as the temperature throughout a building during a user-prescribed fire.

6.6.3 FDS

Fire Dynamic Simulator (FDS) is a large-eddy simulation (LES) code for low-speed flows, with an emphasis on smoke and heat transport from fires. It's one of the most used software to study fire dynamics. The Fire Structure Interface (FSI) module in FDS can be used to impose the gas temperatures from the FDS simulations. It is used to predict the evolving thermal state of the building WTC7 (NIST NCSTAR, 2008).

6.6.4 LS-DYNA

LS-DYNA was capable of explicitly modeling failures, falling debris, and debris impact on other structural components. It could also model nonlinear and dynamic processes, including nonlinear material properties, nonlinear geometric deformations, material failures, contact between the collapsing structural components, and element erosion based on a defined failure criterion. In addition, LS-DYNA had capabilities to include thermal expansion and softening of materials. Therefore, it is used by NIST in their investigation on the World Trade Centre 7 collapse (Figure 6.16).

6.6.5 OpenSees

OpenSees (Open System for Earthquake Engineering Systems) is an open-source software, initially developed by the University of California in collaboration with PEER (the Pacific Earthquake Engineering Research Centre) and NEES (Network for Earthquake Engineering Simulation). OpenSees has a sizeable community of developers and users who are

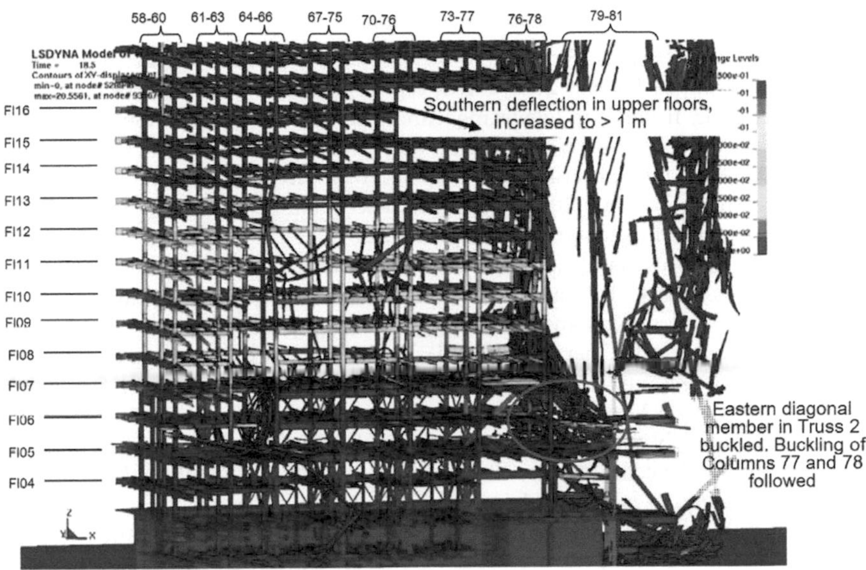

Figure 6.16 Use of LS_DYNA to simulate fire-induced collapse of WTC7. (Reprint from National Institute of Standards and Technology (2005), Federal building and fire safety investigation of the World Trade Center disaster. Final report on the collapse of the World Trade Center Towers. (U.S. Department of Commerce, Washington, DC), NIST NCSTAR 1A, Figures 3–11, p. 40. https://doi.org/10.6028/NIST.NCSTAR.1a.)

devoted to advancing the toolkit available to structural engineers. Although the software was first developed as a tool for assisting in earthquake analysis, it has been extended in recent years and, as such, has emerged as a valuable instrument that structural engineers can use to analyze nonlinear structural responses.

A long-term project was initiated at the University of Edinburgh in 2009 to extend OpenSees' capabilities to the analysis of fire, heat transfer, and thermo-mechanical analyses (Jiang et al., 2012). This involved the establishment of material models at increased temperatures based on the Eurocodes along with thermal load classes. Additionally, to examine thermal effects, temperature-dependent designs have been included for simple element types, including shell and beam elements (Jiang et al., 2012). The material library associated with the early version of OpenSees has also been extended by including temperature-dependent material models for concrete and steel, which are again consistent with the Eurocodes (Jiang et al., 2012). Thermo-mechanical beam-column elements are derived from the base class element, and keep the general interface and data structure from the beam-column elements for ambient temperature use, such as forming the stiffness matrix and residual forces (Jiang, 2012).

REFERENCES

BRE (2004), 'Client report: Results and observations from full-scale fire test at BRE Cardington, 16 January 2003 Client report number 215–741', February 2004. (Accessible from: http://www.mace.manchester.ac.uk/project/research /structures/strucfire/DataBase/TestData/default1.htm)

BS 476-20 (1987), Incorporating Amendment No. 1. Fire tests on building materials and structures —Part 20: Method for determination of the fire resistance of elements of construction (general principles)

Cai, B., Zhang, B. and Fu, F. (2020). Post-fire reliability analysis of concrete beams retrofitted with CFRPs: a new approach. Proceedings of the Institution of Civil Engineers - Structures and Buildings, 173(11), pp. 888–902

Coile, R. V., Caspeele, R., Taerwe, L. (2014), Reliability-based evaluation of the inherent safety presumptions in common fire safety design. *Engineering Structures*, 77, pp. 181–192.

Elhami Khorasani, N., Gernay, T., Garloc, M. (2016), Fire fragility functions for community resilience assessment. *Proceedings of the 9th International Conference on Structures in Fire*, 8–10 June 2016, Princeton University, NJ.

EN 1990:2002+A1 (2005), Eurocode. Basis of structural design, Commission of the European communities.

EN 1991-1-2 (2002), Eurocode 1. Actions on Structures, Part 1–2: General actions. Actions on structures exposed to fire. Commission of the European communities.

EN 1993-1-2 (2004), Eurocode 3. Design of steel structures, Part 1–2: General rules – Structural fire design. BSI, London, UK

EN 1993-1-2 (2005), Eurocode 3. Design of steel structures, Part 1–2: General rules. Structural fire design. Commission of the European communities.

EN 1994-1-2 (2005), Eurocode 4. Design of composite steel and concrete structures, Part 1–2: General rules. Structural fire design. Commission of the European communities.

Fu, F. (2015), *Advanced Modeling Techniques in Structural Design*. John Wiley & Sons, Ltd. ISBN 978-1-118-82543-3.

Fu, F. (2016a), *Structural Analysis and Design to Prevent Disproportionate Collapse*. CRC Press. ISBN 978-1-4987-8820-5.

Fu, F. (2016b), 3D finite element analysis of the whole-building behavior of tall building in fire. *Advances in Computational Design*, 1(4), pp. 329–344

Fu, F. (2020), Fire induced progressive collapse potential assessment of steel framed buildings using machine learning. *Journal of Constructional Steel Research*, 166, pp. 105918–105918.

GB (2010), GB 50010-2010: Code for design of concrete structures. GB, China, Beijing.

GB (2012), GB 50009-2012: Load code for the design of building structures. GB, China, Beijing.

GB (2013), GB 50367-2013: Code for design of strengthening concrete structure. GB, China, Beijing.

Gernay, T., Khorasani, N. E., Garlock, M. (2019), Fire fragility functions for steel frame buildings: Sensitivity analysis and reliability framework. *Fire Technology*, 55(4), pp. 1175–1210. DOI: 10.1007/S10694–018–0764–5

Hopkin, D., Anastasov, S., Swinburne, K., Lay, S., McColl, B., Rush, D., Van Coile, R. (2017), Applicability of ambient temperature reliability targets for appraising structures exposed to fire. *CONFAB 2017 Conference Proceedings.*

Jiang J (2012), Nonlinear thermomechanical analysis of structures using OpenSees. PhD Thesis, University of Edingburgh, Scotland.

NIST (National Institute of Standards and Technology) NCSTAR (National Construction Safety Team) (2008), Federal building and fire safety investigation of the World Trade Center disaster. Final report on the collapse of World Trade Center Building 7. NIST, U.S. Department of Commerce, Gaithersburg, MD.

Novozhilov, V. (2001), Computational fluid dynamics modelling of compartment fire. *Progress in Energy and Combustion Science*, 27(6), pp. 611–666.

Purkiss, J. A. (2007), *Fire Safety Engineering Design of Structures*, Butterworth-Heinemann, Elsevier, Linacre House, Jordan Hill, Oxford, UK

Van Coile, R., Balomenos, G. P., Pandey, M. D., Caspeele, R. (2017), An unbiased method for probabilistic fire safety engineering, requiring a limited number of model evaluations. *Fire Technology*, 53, pp. 1705–1744.

Cai B., Zhao L. L. and Yuan Y. H. (2016) Reliability of bending capacity for corroded reinforced concrete beam strengthened with CFRP. *Concrete* 10, 148–151.

Chapter 7

Preventing fire-induced collapse of tall buildings

7.1 INTRODUCTION

In this chapter, how to design a building to prevent fire-induced collapse will be discussed. The collapse mechanism of a building in fire and methods for mitigating the collapse of a tall building will be introduced, all based on the existing research and fire-induced collapse incidents. As steel-framed buildings are more vulnerable in terms of fire-induced collapse, this type of buildings will be focused in this chapter.

7.2 DESIGN OBJECTIVE AND FUNCTIONAL REQUIREMENT FOR STRUCTURAL STABILITY IN FIRE

Although the first priority of a fire safety design is to save lives of occupants rather than preventing the collapse of buildings (Fu, 2016a), any collapse of the buildings will cause huge economic loss. In addition, collapse of a building can also have a heavy impact on life safety design target. Therefore, building regulation of New Zealand, building Act 1991, specifies a design to have the following requirements:

C4.1 STRUCTURAL STABILITY DURING FIRE OBJECTIVE

- Safeguard people from injury due to loss of structural stability during fire
- Protect household units and other property from damage due to structural instability caused by the fire.

C4.2 FUNCTIONAL REQUIREMENT

- Buildings shall be constructed to maintain its structural stability during fire to
- Allow people adequate time to evacuate safely
- Allow fire service personnel adequate time to undertake rescue and firefighting operations
- Avoid collapse and consequential damage to adjacent house units or other properties.

7.3 IMPORTANCE OF COLLAPSE PREVENTION OF TALL BUILDINGS IN FIRE

From Section 7.2, it can be seen that in a fire safety design, the primary purpose of a building's structural stability is to save the lives of occupants in the event of fire, rather than preventing collapse Currently, the major objective of structural fire design in most design codes is to ensure load-bearing capacity of the building to continue to function until successful evacuation of the occupants, rather than prevent collapse of the building. Therefore, thus far, there is no clear guidance for designing a building to prevent fire-induced collapse across the world.

However, as introduced in Chapter 1, there are several incidents of fire-induced collapse of tall buildings such as World Trade Center (WTC1, WTC2, and WTC7). As explained in Section 7.2, even for the purpose of saving lives of the occupants, preventing or delaying the collapse of a building in fire is also essential in a fire safety design.

7.4 COLLAPSE MECHANISM OF TALL BUILDINGS IN FIRE

All the collapse of a building starts from a local failure of structural members. As introduced in Chapter 2, there are four major failure modes of structural members in fire discovered by Cardington tests:

- Beam buckling and yielding
- Column buckling and yielding
- Connection failure
- Slab failure.

These structural members are not working independently; they have impacts on each other. Under thermal expansion and subsequent contraction in the cooling stage, the interaction between the structural members causes extra stress and deformation to them, which makes it difficult to predict the actual failure mode of the whole building. However, as introduced in Chapter 2, it is found from WTC1 and WTC7 collapse incidents that the column buckling is the key reason for the collapse of these two buildings. The column buckling will trigger the failure of the floor above. The floor failure in both WTC1 and WTC7 caused further failure of surrounding columns in the horizontal direction and triggered progressive failure of the whole building. Therefore, it is worth investigating the response of individual structural members first.

7.4.1 Factors affecting thermal response and failure mechanism of individual members

The thermal behavior of structural members in fire is a complex problem, and the response of individual members in a tall building subjected to fire loading is affected by the following parameters:

1. Temperature profile of the member
2. Degree of thermal restraint offered by the surrounding members
3. Degradation in material properties with increasing temperature
4. Capacity of deployment of alternative load-carrying paths for adjacent members.

To better understand the behavior of the structural member in fire, a 3D finite element model of a typical single-storey composite frame is set up in Abaqus® by the author, as shown in Figure 7.1. To facilitate further discussion on the modeling result, this model is designated as Model 1 in this chapter.

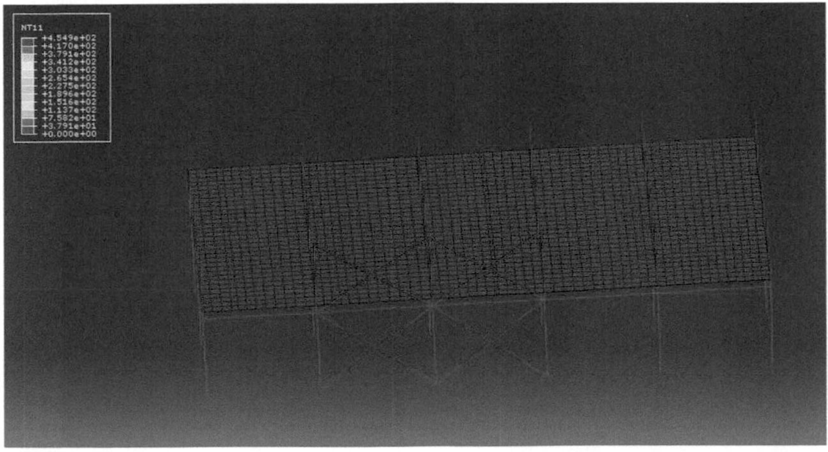

Figure 7.1 Model 1 of a typical composite floor in Abaqus®. (Abaqus® screenshot reprinted with permission from Dassault Systèmes.)

In the analysis, parametric fire temperature is used for gas temperature, Short-hot fire scenario is adopted. Figure 7.1 shows the temperature distribution of slabs, beams, and columns after analysis. The edge columns and beams and columns are fire protected. Therefore, it can be seen that the temperature of the slabs is higher than that of beams and columns due to their fire protection.

7.4.2 Behavior and failure mechanism of steel beams in fire

7.4.2.1 *Local buckling of beams in connection area*

In the event of fire, the steel beams experience expansion at the heating stage, but this expansion is restrained by the connections and columns which they are connected to. To clearly understand the behavior of the beams in a fire condition, Model 1 is used here to demonstrate the change of internal force in the beams. As shown in Figure 7.2, the axial force of the beam first increases due to thermal expansion; however, due to the restrain of the connections and columns, the tensile force starts to drop at around 300 s, as the constrain causes the compression force in the beams.

In Cardington tests, all primary beams showed signs of local buckling near to the connections. All the secondary beams also exhibited local buckling at the connection zone as well. However, local buckling did not occur where beams were supported on edge beams. This is due to the buckling stresses being relieved by rotation of the edge beam at the connection. In most of the beams, lower flange buckling was observed. The reason for local buckling at connection zone is due to the constraint of the connection to the beam during the expansion in the heating stage.

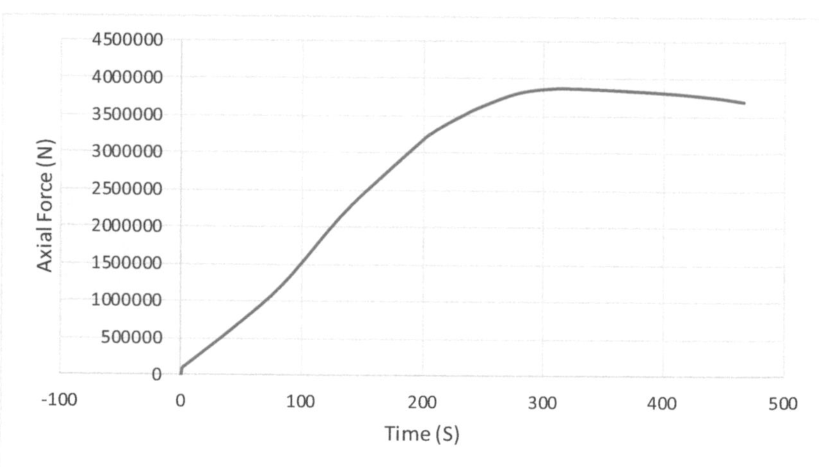

Figure 7.2 Tensile axial forces of beams.

This constraint causes plastic compression deformation of the beam especially in the lower flange. This is clearly demonstrated in in Figure 7.3. It can be seen that, the plastic strain developed in the beam at connection location during the fire in Model 1. It can be seen that at around 200 s, plastic strain starts to develop.

7.4.2.2 *Excessive deflection*

In all Cardington tests, the unprotected steel beams were severely deformed, at a beam temperature of 500°C–600°C. As shown in Figure 7.4, the deflection of the beam increases when the temperature increases.

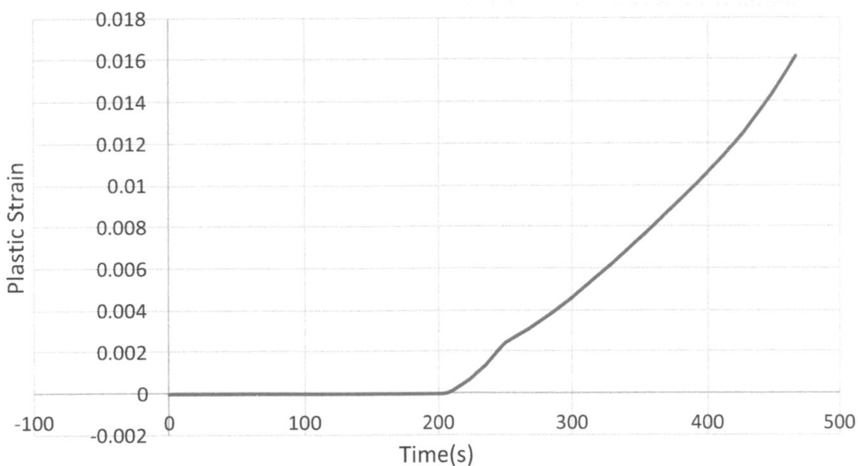

Figure 7.3 Plastic stain development of the beam at connection location.

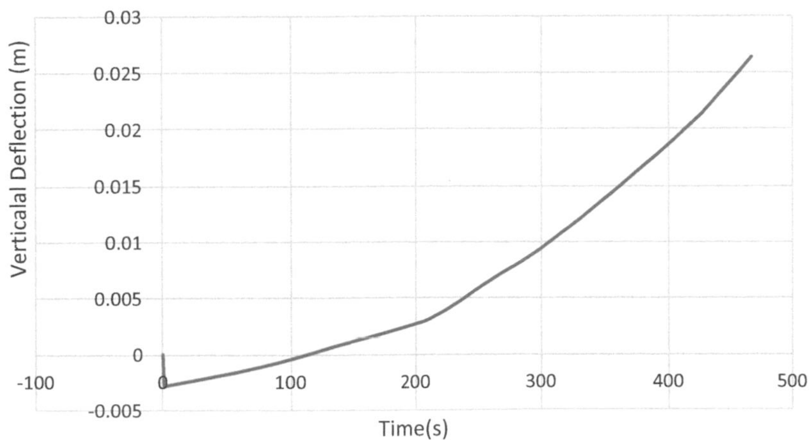

Figure 7.4 Vertical deflection of beam at middle span.

7.4.3 Behavior and failure mechanism of slabs in fire

Figure 7.5 shows the deflected shape of a typical steel composite floor in fire. As introduced in Chapter 3, Section 3.4.3, the slab was heated at a much slower rate than the steel beam due to its greater thermal inertia. It has a greater thermal gradient compared to steel beams, with a lower mean slab temperature. Initially, the slab response was driven largely by the behavior of the rapidly expanding steel beams. This was due to their large compressive forces to the corresponding post-buckling deflections of the composite beam and to the consequent large moments due to a P-δ effect. However, after the steel beams could no longer transfer a force to the slab, the stress state within the slab itself drove the response of the floor system as a whole (O'Connor et al., 2003).

7.4.3.1 Membrane actions of slabs

At ambient temperature, the failure mode of slabs follows patterns of yield lines. However, based on the observations of the BRE Cardington tests and other fire tests on tensile membrane action, mode of failure and the design method are suggested by Bailey and Moore (2000a,b) as introduced in Chapter 5.

In Cardington tests, despite failures of steel frame, the floor slab continued to resist loading due to membrane action developed in the slab. Damage to the floor slab steel mesh was found. In all tests, concrete cracks were observed along compartment boundaries or the main primary and secondary beams. The most extensive damage to the concrete slab was observed around the central column. This damage occurred at the cooling stage due

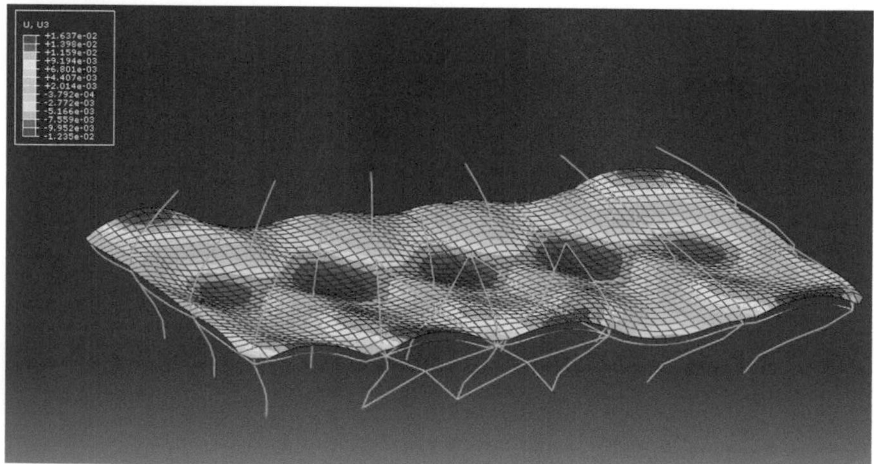

Figure 7.5 Abaqus® model showing deflected shape of a typical steel composite floor. (Abaqus® screenshot reprinted with permission from Dassault Systèmes.)

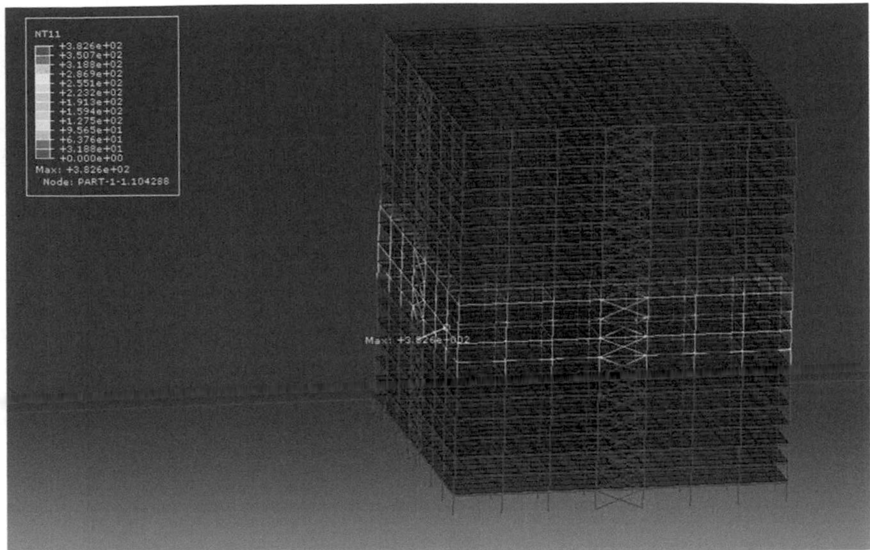

Figure 7.6 Model 2, a tall building under long-cool fire in storeys 9–11. (Abaqus® screenshot reprinted with permission from Dassault Systèmes.)

to the high forces generated by differential cooling period between concrete and steel structures. To tackle this failure mode, sufficient anchorage of reinforcing mesh at the boundary is essential in keeping the integrity of the compartment as well as resisting potential collapse of the buildings.

Figure 7.6 shows an Abaqus® finite element model set up by the author to simulate a tall building under long-cool fire. It is denoted as Model 2 in this chapter. The model replicates a 20-storey steel-framed tall building with cross-bracing in four sides of the building as major lateral stability system. It has 7×7 bays at each direction. The fire was set in storeys 9–11.

Figure 7.7 is plastic strain distribution in the floor plate at storey 11. It can be seen that plastic strain developed at each bay following the pattern of membrane action. The floor slabs resisted the loads by tensile membrane action and formed a tensile zone in the middle of the fire-exposed slab, surrounded by a compression ring.

It is recommended that fire protection will provide the necessary perimeter vertical support along the slab panel boundaries to ensure membrane action and, therefore, resist the collapse of the building.

7.4.3.2 Effect of different fire scenarios in composite slabs

As discussed in Section 3.4.4 of Chapter 3, the research by Lamont et al. (2004) discovered that the stress state of the composite beams is significantly affected by the heating regime. The most detrimental fire scenario

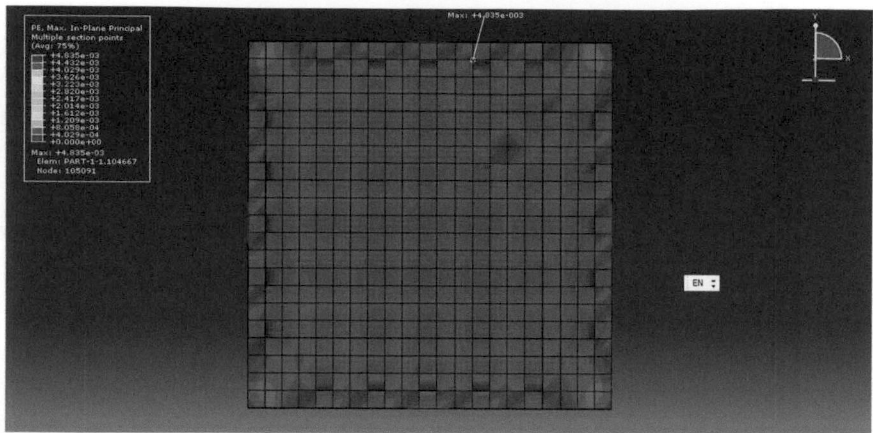

Figure 7.7 Principle plastic strain distribution of floor plate at storey 11. (Abaqus® screenshot reprinted with permission from Dassault Systèmes.)

for composite beams is the "short-hot" fire. Because the temperature gradient is high, greater thermal bowing effect, causing a greater tension in the composite section and large tensile strains, is observed in the slabs.

Long-cool fire results in higher temperatures in the concrete and the protected steel. This results in greater displacements in the protected structural elements but much later in terms of real time. However, because the concrete slab achieves higher temperatures, there is much less tension in the slab with growing compression towards the end of heating.

7.4.3.3 Other research in composite slabs in fire

Wong and Burgess (2013) found that the vertical supports provided by protected beams along the edge of the slab are important in the development of the tensile membrane actions.

Lin et al. (2015) investigated the effect of protected beams on the fire resistance of composite buildings. The results showed that non-fixed vertical support significantly reduced the development of tensile membrane action. In comparison to the case with fixed support, tensile membrane action was fully mobilized.

Jiang and Li (2014) studied parameters affecting tensile membrane action of reinforced concrete floors in fire. It was found that failure modes of the slab depend on reinforcement layout, aspect ratio, and boundary condition.

Nguyen and Tan (2017) conducted experiments on three one-quarter scale composite slabs with different bending stiffnesses of protected edge beams under fire conditions. The results showed that an increase in the edge beam-bending stiffness initially reduced the deflection. At higher temperature, the effect of greater stiffness of the edge beams was negligible.

All of the above studies confirmed that protected edge beams have a significant effect on the fire resistance of the structure.

7.4.4 Behavior and failure mechanism of steel column in fire

7.4.4.1 *Change of column force in fire*

Figures 7.8–7.10 show the development of axial compression force of columns at different locations in Model 2. It can be seen that due to thermal expansion, the axial compression force in corner and edge column shows

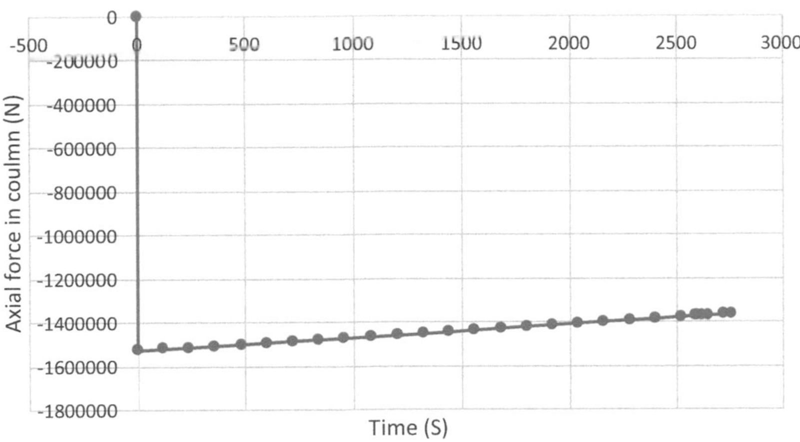

Figure 7.8 Compression axial forces of a perimeter column at level 10 during fire.

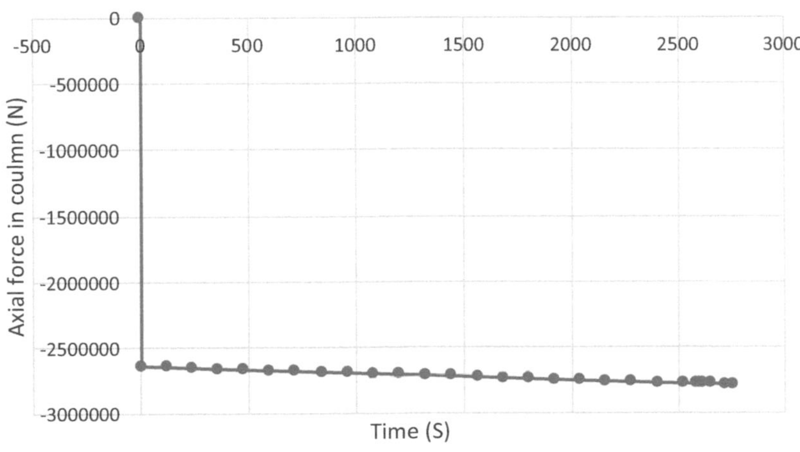

Figure 7.9 Compression axial forces of an internal column at level 10 during fire.

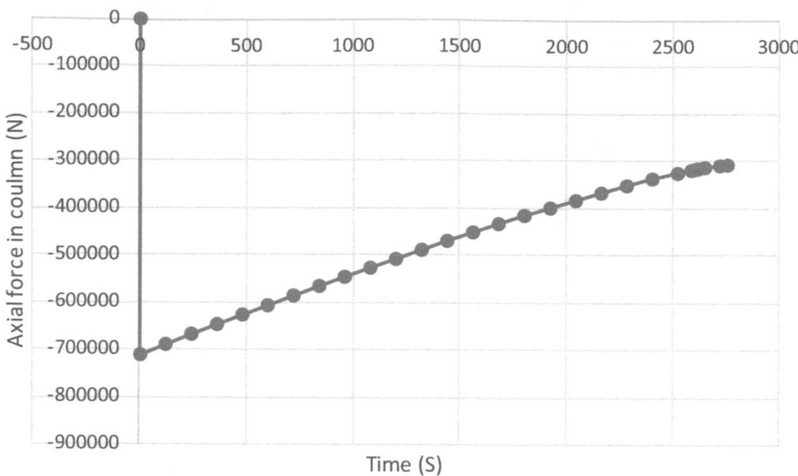

Figure 7.10 Compression axial forces of a corner column at level 10 during fire.

certain percentage of decrease when the temperature of fire rises. The corner columns show a greater drop of compression force than the edge columns. However, the axial forces in internal columns show an increase. The reason for these force distributions can be explained by Figure 7.11, which shows the overall deformed shape of the entire building in fire. It can be seen that due to the thermal expansion, the edge and corner columns are

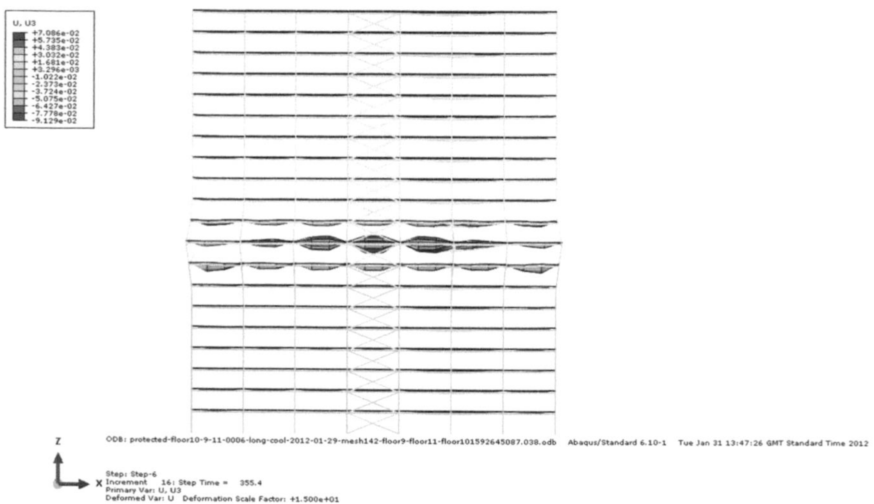

Figure 7.11 Deform shape of Model 2 in fire (deformation amplified 15 times). (Abaqus® screenshot reprinted with permission from Dassault Systèmes.)

deforming outward, therefore making their force reduce and increasing the column force in internal columns.

7.4.4.2 Out plane bending of columns

To better understand the behavior of the columns, the Cardington test is simulated using ANSYS as shown in Figure 7.12. Fire was set at storey 4 in the corner to simulate Conner fire test in the Cardington test (as shown in Figure 7.12, which shows temperature contour of the building elements). This is different to the fire scenarios in Model 2, where all three floors are set to fire.

Figure 7.13 shows the moment of the columns on the third floor under fire loading. A huge bending moment was observed. Due to the effects of material degradation, the bearing capacity of these columns under fire is significantly reduced, so the buckling load reduces. As we know, buckling of the columns is one of the major mechanisms for collapse of the building under fire. However, extra bending moment due to fire will further decrease the buckling load of the columns making them more vulnerable. It is also noticed that edge and corner columns that are not in direct contact with the fire exhibit bending moment around 70% of those in contact with fire, while inner columns exhibit around 30% of the bending moment of those in contact with fire. This indicates that the fire will significantly increase the bending moment of the columns during the fire.

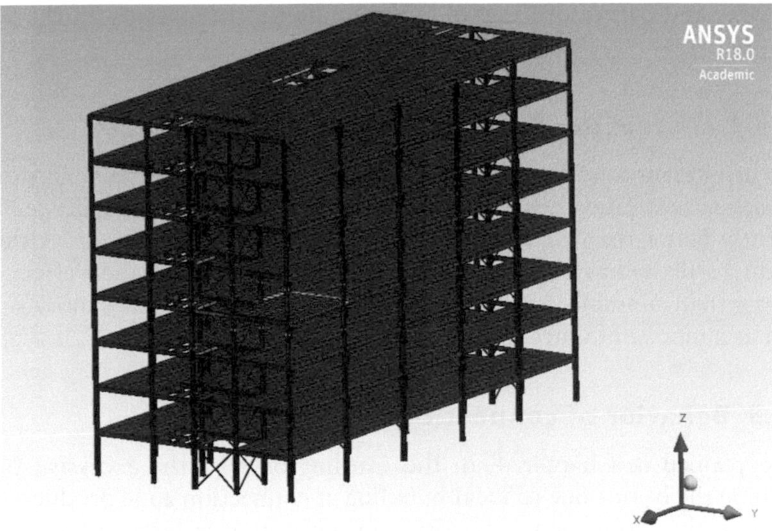

Figure 7.12 Model of Cardington fire test in ANSYS. (Screenshot reprinted with permission from ANSYS, Inc.)

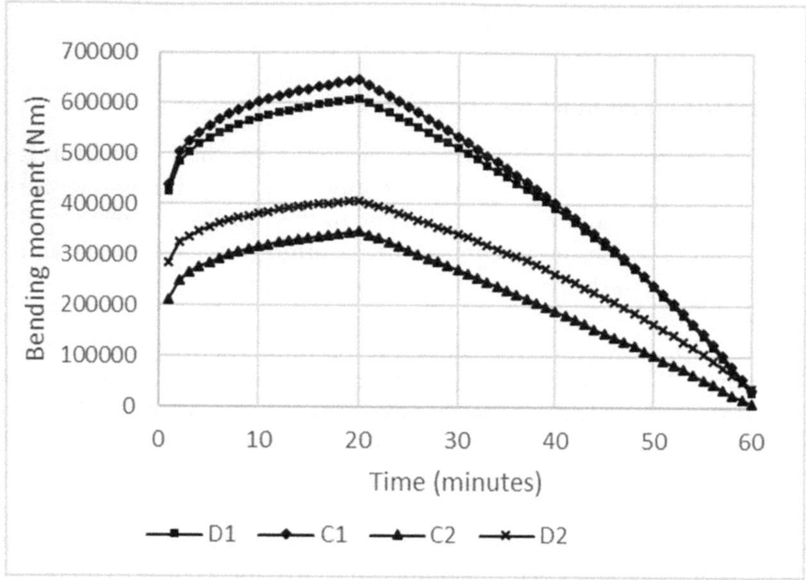

Figure 7.13 Bending moment of columns under fire loading.

Therefore, it can be concluded that fire will further decrease the buckling load of the column, and the corner and edge columns are more vulnerable during the fire. Fire protection should be made to columns especially at the edge and corner of the building to enhance the stability of the whole building.

7.4.4.3 *Effect of the slenderness ratios*

The investigation by Burgess et al. (1992) found that columns with different slenderness ratios behave differently in fire. Stocky and slender columns perform better than those with intermediate slenderness ratios. Although residual stresses have been shown to influence the failure, their effect is no greater than at ambient temperature. Local buckling becomes more significant as a mode of failure in fire.

7.4.5 Behavior of connections

As explained in Chapter 4, in the Cardington test, the excessive plastic strain in the beams due to local buckling at connection zone produces huge tensile at the cooling stage, resulting in connection failures in both partial depth endplate and fin plate connections.

7.4.6 Behavior and failure mechanism of concrete column in fire

Research by Dimia et al. (2011) shows that a failure during the cooling phase of a fire is more dangerous, that a failure of the structure is still possible when the fire has been completely extinguished. It has been shown that the most critical situations with respect to delayed failure arise for short-hot fires and for columns with low slenderness (short length and/or massive section). Rapid cooling of the gas temperature increases the probability for the column to survive the fire.

7.5 WHOLE-BUILDING BEHAVIOR OF TALL BUILDINGS IN FIRE

7.5.1 Research of Fu (2016b)

Fu (2016b) simulated the whole-building behavior of a tall building using 3D finite element method. Based on the modeling observations and analysis of data, the following findings were made:

- 'Strong column and weak beam' is a design principle widely used to prevent early collapse of the buildings, which enables the beams to fail earlier than columns. However, under fire, due to the thermal expansion and the strength degradation of the material, the columns are prone to fail earlier than the beams, so this design principle is not applicable. Therefore, the effective way to prevent the building collapse is to prevent the early failure of the columns.
- Due to thermal expansion, the corner slabs are more vulnerable to fail than the slab at other locations.
- Due to the global deformation of the whole floor plate and the restrain from the supporting steel beams, tensile membrane and compressive membrane are developed in the slab. This membrane action will enhance the load-carrying capacity of the slab. However, it is also noticed that with the increasing of temperature, the axial force of the slab dropped, which is due to the large deformation of the supporting beams.

7.5.2 Twin Towers (WTC1 and WTC2)

The investigation of NIST (2005) found that the aircraft hit the towers at a high speed impacting between the 93rd and 99th floors, and did considerable damage to principal structural components (core columns, floors, and perimeter columns) that were directly impacted by the aircraft or associated debris. However, the towers withstood the impacts and would have remained standing were it not for the dislodged insulation (fireproofing) and the subsequent multi-floor fires.

7.5.2.1 Structural framing for WTC1

As shown in Figure 7.14, rather than using the traditional metal decking composite slab in the floor system, WTC1 adopted a so-called composite truss floor system due to its extra-long span of the floor. As explained in Chapter 2, the bowing of the composite floor becomes one of the key reasons for the collapse of WTC1.

Two-dimensional FEM models were built by Flint et al. (2007) and Usmani et al. (2003) to investigate the collapse of the World Trade Centre towers. They found that tensile membrane action occurred by floor deflection caused by thermal expansion, with inward pulling of the exterior columns, which led to the formation of plastic hinges in those columns at floor levels.

7.5.2.2 Reason for the collapse of WTC1

As shown in Figure 2.10, Chapter 2, WTC1 and WTC2 used so-called perimeter frame-tube system with excellent robustness, and the large size of the buildings helped the towers withstand the impact. The structural system redistributed loads from places of aircraft impact, avoiding larger-scale damage upon impact. The hat truss was intended to support a television antenna and prevented earlier collapse of the building core. In each tower,

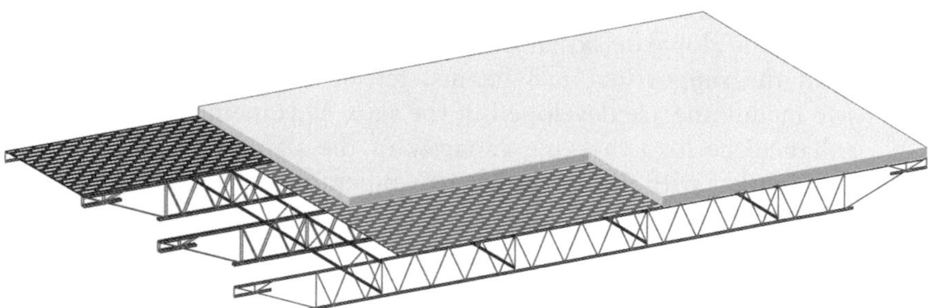

Figure 7.14 Composite truss floor system of Word Trade Center. (Reprinted from National Institute of Standards and Technology (2005), Federal building and fire safety investigation of the World Trade Center disaster, final report on the collapse of the World Trade Center Towers. (U.S. Department of Commerce, Washington, DC), NIST NCSTAR 1, Figures 1–6, p. 10. https://doi.org/10.6028/NIST.NCSTAR.1.)

a different combination of impact damage and heat-weakened structural components contributed to the abrupt structural collapse.

As explained in Chapter 2, in WTC1, the fire weakened the core columns and caused the floors on the south side of the building to sag. The floors pulled the heated south perimeter columns inward, reducing their capacity to support the building above. Their neighboring columns quickly became overloaded as columns on the south wall buckled. The top section of the building tilted to the south and began its descent.

7.5.3 WTC7

In WTC7, according to NIST NSCTAR1A (2008), the fire lasted only on floors 7–9 and 11–17 until the building collapsed.

7.5.3.1 Structural framing for WTC7

As shown in Figure 7.15, the plan layout shows the framing of WTC7; it is primarily composed of five major frames (columns 1–42, 58–79, 59–80, 60–81, and 15–28 in the horizontal direction).

As shown in Figure 7.16, there are several two-storey high trusses between floors 5 and 7. The cantilever transfer girders are also used in the vertical direction making the whole structural framing quite unique.

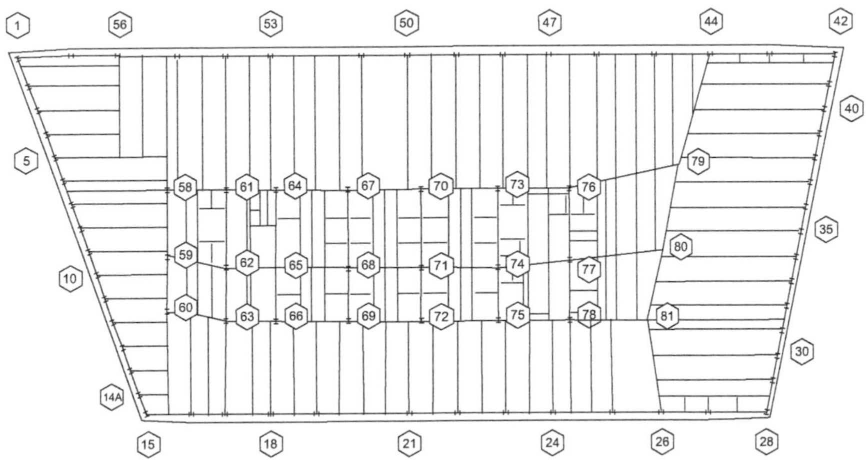

Figure 7.15 The floor layout of WTC7. (Adapted from National Institute of Standards and Technology (2005), Federal building and fire safety investigation of the World Trade Center disaster, final report on the collapse of the World Trade Center Towers. (U.S. Department of Commerce, Washington, DC), NIST NCSTAR 1A, Figures 1–5, p. 6. https://doi.org/10.6028/NIST.NCSTAR.1a.)

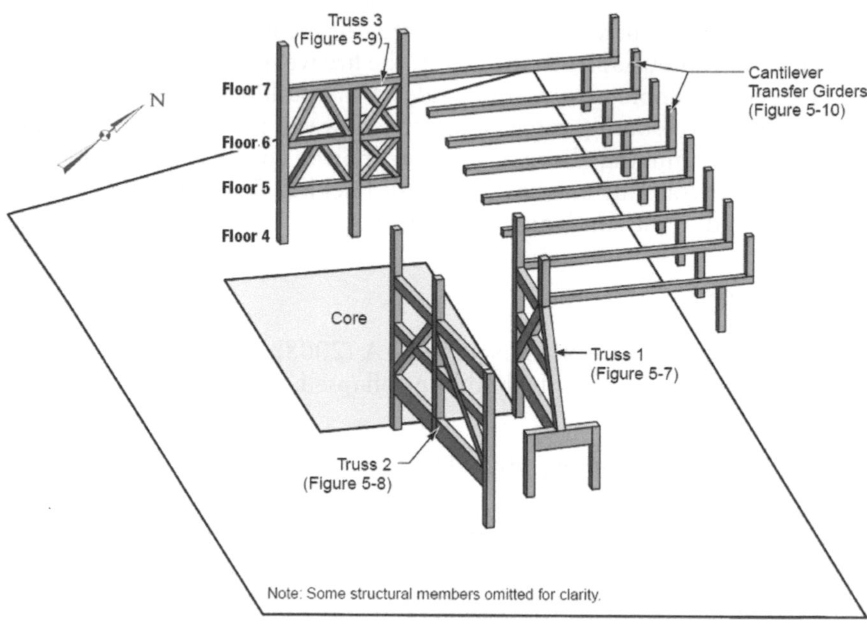

Figure 7.16 Key structural system of WTC7. (Reprint from National Institute of Standards and Technology (2005), Federal building and fire safety investigation of the World Trade Center disaster, final report on the collapse of the World Trade Center Towers. (U.S. Department of Commerce, Washington, DC), NIST NCSTAR 1A, Figures 1–6, p. 6. https://doi.org/10.6028/NIST.NCSTAR.1a.)

NIST (2008) made simulation of the collapse of the building using LS-DYDA. As shown in Figure 7.17, the collapse is triggered by the buckling of column 76 between levels 7 and 9. The splice of column 76 failed in bending, led to floor failure and bending failure of adjacent columns, causing a progressive failure. It is worth noting that the trusses in levels 5 and 7, which are in certain degree similar to the composite truss floor system in WTC1, have a larger stiffness; therefore, when a column fails, it pulls the adjacent column, causing continuous failure of all columns.

7.5.3.2 Reason for the collapse of WTC7

To clearly investigate the collapse mechanism of WTC7 in fire, it was simulated by the author using Abaqus® as shown in Figure 7.18. It can be seen that the exterior columns buckled at the lower floors (between floors 7 and 14) due to load redistribution to the exterior columns from the building core. The interior columns buckled due to fire. The entire building above

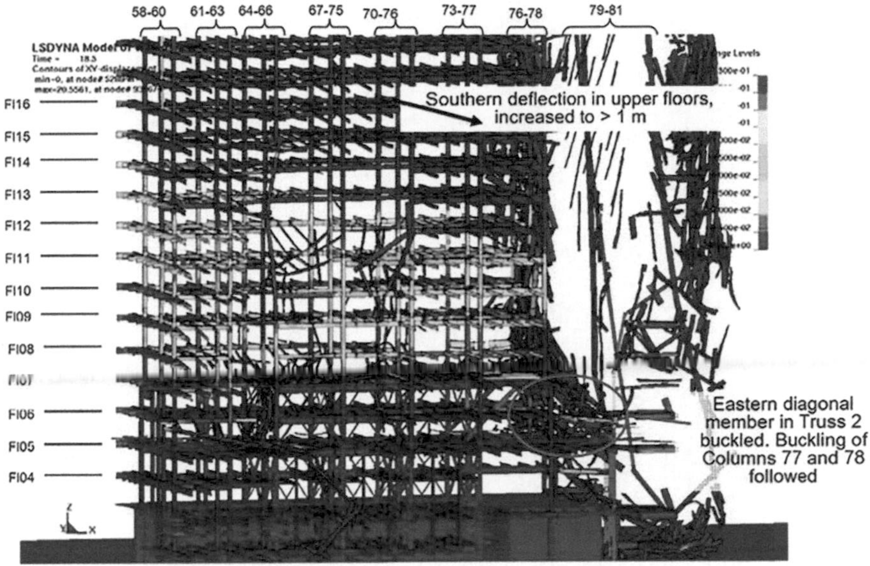

Figure 7.17 Vertical progression of failure simulated by LS_DYNA. (Reprint from National Institute of Standards and Technology (2005), Federal building and fire safety investigation of the World Trade Center disaster, final report on the collapse of the World Trade Center Towers. (U.S. Department of Commerce, Washington, DC), NIST NCSTAR 1A, Figures 3–10, p. 40. https://doi.org/10.6028/NIST.NCSTAR.1a.)

the buckled-column region then moved downward in a single unit, and as observed, the global collapse was triggered.

The investigation of NIST NCSTAR1A (2008) also shows that

- Long-span floor system experiences significant thermal expansion and sagging effects.
- Connection designs (especially shear connections) cannot accommodate thermal effects.
- Floor framing induces asymmetric thermally induced (i.e., net lateral) forces on girders.
- Shear studs fail due to differential thermal expansion in composite floor systems.

7.5.4 Cardington test

In the Cardington test, the structural integrity of the composite frame was maintained during both the heating and cooling phases. This is due to several reasons.

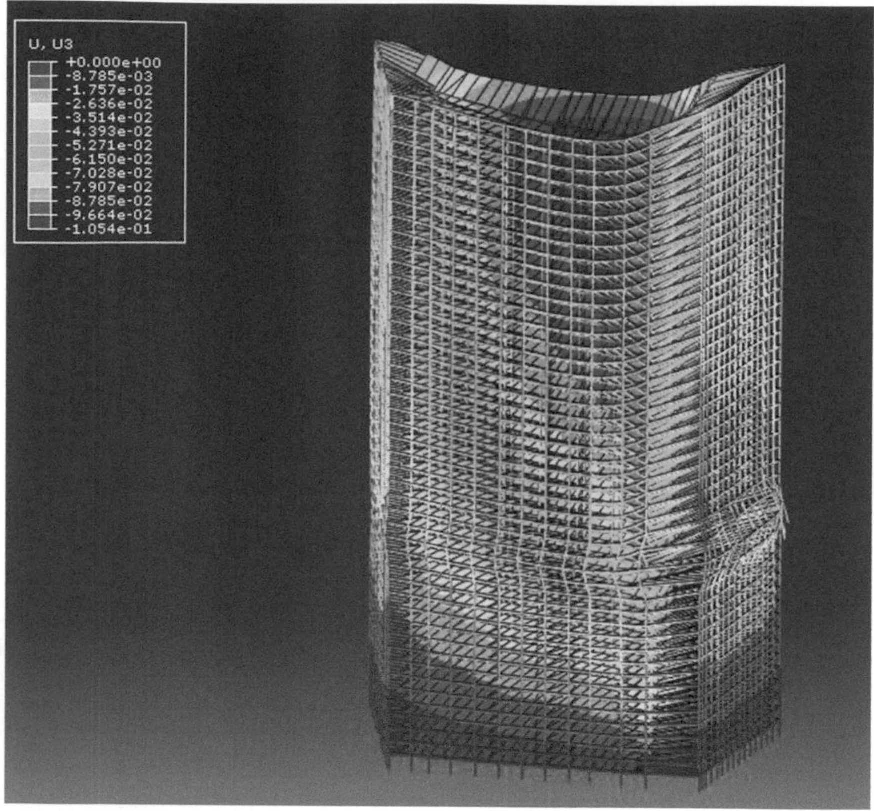

Figure 7.18 Abaqus® model showing the collapse of WTC7 building. (Abaqus® screenshot reprinted with permission from Dassault Systèmes.)

7.5.4.1 Severity of the fire

As introduced in Chapter 2, in Cardington, the fire was only set in certain area, rather than spread into the whole floor plate or floor above or underneath, making less server fire than WTC1 and WTC7.

7.5.4.2 Structural framing

As shown in Figure 2.16, the structural framing of Cardington is more regular compared to WTC1 and WTC7 which are kind of special. The structural framing of both WTC1 and WTC7 has been introduced in the preceding sections, and it can be seen that compared to WTC1 and WTC7, the spacing of the columns in Cardington is much smaller, and therefore less span of the slabs, making the membrane action, can be easily developed which helps to resist the collapse of the building. However, due to the unique flooring system of WTC1 and WTC7, membrane action did not stop the collapse of the building. And as it has been explained, the long-span slabs in both WTC1 and WTC7 partially contributed to the collapse of the two buildings.

7.5.5 Other research in whole building behavior

A 2D finite element model for a steel moment frame with 38 storeys was analyzed by Garlock and Quiel (2007) using real fire scenarios, including a temperature–time curve consisting of heating and cooling stages. Exposure to fire was restricted to a single exterior frame bay, with vertical diffusion of the fire from the 22nd to 30th floors. According to the observations, lateral deformation of the perimeter columns occurred as a result of the fact that the heated beams underwent expansion during the stage of heating. The risk of structural collapse was increased with the formation of plastic hinges that occurred upon the beams and columns reaching the limit state under the joint action of axial forces and bending moments. Another important observation was that the perimeter column stability was dependent on the beams that braced those columns, they contributed more significantly to protection against fire.

Lange et al. (2012) used two-dimensional finite element models to analyze a 12-storey composite steel frame with interior core and perimeter moment frame in fire. It is found that as the heated floors became thermally expanded, the perimeter columns were pushed outward at first, while the major deflection and catenary action of the floors subsequently caused the columns to be pulled inward. In the end, the building collapsed gradually due to the buckling of the heated perimeter columns.

The progressive collapse of a typical super-tall RC frame-core tube building exposed to extreme fires was simulated by Lu et al. (2017). It is found that

1. The progressive collapse may originate in the fire-affected areas. However, due to the large redundancy of the super-tall RC frame-core tube building, the progressive collapse of the residual structure outside of the fire-affected areas can be prevented via alternative load paths.
2. The progressive collapse of the super-tall building was triggered by the flexural failure of the peripheral columns, due to the outward push by the thermal expansion of the upper floors and the inward contraction of the lower floors.
3. The more stories subjected to fire, the greater the possibility of progressive collapse.

Three-dimensional FEM models were employed by Agarwal and Varma (2014) to investigate a steel moment frame building. Similar to the research results of Fu (2016b), they also found that columns had the highest probability of undergoing failure first, if the fire safety level was identical for every structural element. When the gravity columns fail, load redistribution to neighboring columns is induced by catenary and flexural action.

7.6 OVERALL BUILDING STABILITY SYSTEM DESIGN FOR FIRE

7.6.1 Bracing system

In tall buildings, all parts of the bracing system should be fire protected to ensure appropriate fire resistance. To ensure the overall stability of the buildings, another strategy is to place the bracing inside a protected shaft such as stairwell. However, it should ensure that the lift shaft has sufficient fire resistance.

7.6.2 Core wall design

In the history of tall building design, steel core was dominant in most of the steel-framed tall buildings, such as WTC1 and WTC2. The lessons from WTC collapse show that sole steel material is not advised. Therefore, since the 9/11, the steel core is seldom used in tall buildings. All the tall buildings have started using concrete core as it is introduced.

7.7 METHODS FOR MITIGATING COLLAPSE OF BUILDINGS IN FIRE

From the above investigation, it can be seen that the structural framing plays an important role in building collapse when fire starts. Special attentions should be paid to the columns during the design, especially to the edge and corner columns.

Therefore, the following measures can be recommended to prevent the collapse of tall buildings in fire:

1. The collapse of WTC1 and WTC7 and the research of Fu (2016b) and other researchers all show the important role of columns in the collapse of the buildings. And the collapse of all these three buildings was triggered by the edge columns. The research of Fu (2016b) and Agarwal and Varma (2014) shows the high likelihood of early failure of columns in building fire. Therefore, prevent early failure of the columns of tall building in fire is an effective way to prevent fire induced collapse. Apart from fire protections, this can be achieved through increasing the out-plan moment capacity and shear capacity of the column, such as increasing the flange and web thickness.
2. In designing the protection regimes, special attention should be made to corner and edge columns. They are experiencing large out-plane bending due to the thermal effect, as it is explained in the simulation results of both Model 2 and Cardington tests. It is suggested to increase the thickness of the fire protection.

3. Most of the connections failed due to the thermal expansion and subsequent contraction of the steel beams at the cooling stage. A robust connection with flexible deformation is needed for the sake of preventing collapse of the buildings.
4. Protected edge beams have a significant effect on the fire resistance of the structure.
5. Structural systems are also key to preventing progressive collapse. A good structural framing can provide better resistance to the effects of thermal expansion on the structural system. The lessons from WTC1 and WTC7 show that avoiding long spanning floor system and increasing the redundancy of the structure can reduce the chance of collapse. In addition, using a concrete core rather than a steel core as a major lateral stability system would greatly reduce the chance of collapse.
6. Good fire safety design in turn will reduce the likelihood of building collapse. It includes the following:
   - Good thermal insulation to limit heating of structural steel and to minimize both thermal expansion and weakening effects is important to prevent the collapse of a building currently.
   - Sufficient compartmentation to limit the spread of fires.
   - Automatic fire sprinkler systems with independent and reliable sources for the primary and secondary water supplies.
   - Thermally resistant window assemblies which limit breakage, reduce air supply, and retard fire growth.
7. As discussed in Chapter 2, NIST NCSTAR (2008) recommends a performance-based design for fire resistance that can effectively prevent collapse of a building in fire. In prescriptive design methods, the fire rating did not prevent the collapse of building. Therefore, using performance-based design is essential in preventing the collapse of the tall buildings in fire.

REFERENCES

Agarwal, A., Varma, A. H. (2014), Fire induced progressive collapse of steel building structures: The role of interior gravity columns. *Engineering Structures*, 58, pp. 129–140.

Bailey, C. G., Moore, D. B. (2000a), The structural behaviour of steel frames with composite floor slabs subject to fire: Part 1: Theory. *The Structural Engineer*, 78(11), pp. 19–27.

Bailey, C. G., Moore, D. B. (2000b), The structural behaviour of steel frames with composite floor slabs subject to fire: Part 2: Design. *The Structural Engineer*, 78(11), pp. 28–33.

Burgess, I. W., Olawale, A. O., Plank, R. J. (1992), Failure of steel columns in fire. *Fire Safety Journal*, 18, pp. 183–201.

Dimia, M., Guenfoud, M., Gernay, T., Franssen, J. (2011), Collapse of concrete columns during and after the cooling phase of a fire. *Journal of Fire Protection Engineering*, 21(4), pp. 245–263.

Flint, G., Usmani, A., Lamont, S. (2007), Structural response of tall buildings to multiple floor fires. *Journal of Structural Engineering*, 133(12), pp. 1719–1732.

Fu, F. (2016a), Structural Analysis and Design to Prevent Disproportionate Collapse. CRC Press. ISBN 978-1-4987-8820-5.

Fu, F. (2016b), 3D finite element analysis of the whole-building behavior of tall building in fire. *Advances in Computational Design*, 1(4), pp. 329–344.

Garlock, M. E. M., Quiel, S. E. (2007), The behavior of steel perimeter columns in a high-rise building under fire. *Engineering Journal, AISC*, 44(4), pp. 359–372.

Jiang, J., Li, G.-Q., Usmani, A. (2014), Progressive collapse mechanisms of steel frames exposed to fire. *Advances in Structural Engineering*, 17(3), pp. 381–398.

Lamont, S., Usmani, A. S., Gilliec, M., (2004), Behaviour of a small composite steel frame structure in a "long-cool" and a "short-hot" fire. *Safety Journal*, 39, pp. 327–357.

Lange, D., Röben, C., Usmani, A. (2012), Tall building collapse mechanisms initiated by fire: Mechanisms and design methodology. *Engineering Structures*, 36, pp. 90–103.

Lu Xinzheng, Li Yi, Guan Hong, Ying Mingjian (2017), Progressive collapse analysis of a typical super-tall reinforced concrete frame-core tube building exposed to extreme fires, Fire Technology, 53, pp. 107–133.

O'Connor, M. A., Kirby, B. R., Martin, D. M. (2003, January), Behaviour of a multistorey composite steel framed building in fire. *The Structural Engineering*, pp. 27–36.

Nguyen, T., Tan, K. (2017), Behaviour of composite floors with different sizes of edge beams in fire. *Journal of Constructional Steel Research*, 129, pp. 28–41.

NIST NCSTAR 1A (2008), Final report on the collapse of world trade center building 7, National Institute of Standards and Technology, US Department of Commerce.

NIST NCSTAR (2005, December), Federal building and fire safety investigation of the World Trade Center disaster, final report of the National Construction Safety Team on the collapses of the World Trade Center Towers.

NZS 3404 Parts 1 and 2 (1997), Steel Structures Standard.

Usmani, A. S., Chung Y. C., Torero, J. L. (2003), How did the WTC towers collapse: A new theory. *Fire Safety Journal*, 38(6), pp. 501–533.

Wong, B., Burgess, I. (2013), The influence of tensile membrane action on fire-exposed composite concrete floor-steel beams with web-openings. *Procedia Engineering*, 62, pp. 710–716.

Lin, S., Huang, Z., and Fan, M. (2015). The effects of protected beams and their connections on the fire resistance of composite buildings. *Fire Safety Journal*, 78, pp.31–43.

Chapter 8

New technologies and machine learning in fire safety design

8.1 INTRODUCTION

There are many new technologies developed recently which can be used for fire safety design, such as PAVA system, IOT, and smart building management system (BMS). They will be introduced in this chapter. Machine learning (ML) is a new technology that is gradually replacing human beings in most of the disciplines. Their applications in construction industry are still restricted. However, it will have a great impact in all areas of the global economy in the future. It will change the construction industry fundamentally. Therefore, it is beneficial to the engineers to have some basic ML knowledge. For this purpose, some pilot studies of using machine leaning in fire safety design will also be introduced in this chapter.

8.2 NEW TECHNOLOGIES IN FIRE SAFETY

With the fast development of the technology, some new technologies such as PAVA, IOT, and smart BMS have been developed. They have the great potential to enhance current fire safety design measures and bring the fire safety design to a new level.

8.2.1 PAVA alarm systems

Personal Announcement Voice Activated (PAVA) alarm systems integrate alarm detection systems to controlled evacuation of buildings by means of clear pre-recorded spoken messages rather than bells or sounders. In this way, the system automatically broadcasts specific messages to alert occupants in case of a danger and direct them to the nearest safe exits or refuge areas. Traditional bells and sounders only give a warning, but they do not indicate the nature of the emergency. This may leave people uncertain, and often such alarm signals are ignored—potentially with fatal consequences. A voice alarm system provides clear easily understood instructions via pre-recorded messages, ensuring that even untrained personnel can be evacuated speedily and efficiently. Therefore, as it reduces the reaction time very effectively, it is especially effective for complex buildings.

Phased evacuation can be assisted by PVAV using a combination of clear pre-recorded messages and live announcements to enable occupants in selected areas to be evacuated in turn. The voice alarm system works automatically, with all controls easily overridden by fire officers or building control when needed.

8.2.2 IOT in fire safety

Internet of Things (IOT) has been gradually used in fire safety design recently. The IOT usually refers to the idea that everyday objects could be connected to the internet.

8.2.2.1 Fire safety sensors and BMS

IOT sensors are now used to refer to any small, internet-connected devices which record specific data and transmit this to a central location or device to be interpreted. These sensors might record audio, video, temperature data, location data, and much more besides.

Smart buildings are properties which are controlled in part by autonomous computer software, known as BMS. These sensors can maintain a specific temperature in different rooms, turn lights on and off, and perform other tasks which benefit from external data.

Special heatproof sensors could detect the temperature of fires, giving firefighters a clue of intensity of fires, allowing them to alter their equipment and approach. IOT sensors could show not only the location of a fire started but also the speed and direction of its spread. All of this information could be transmitted automatically to fire crews, even happening alongside the emergency call.

The emergency calls will be automated by the BMS, with the system forwarding vital data to the local fire department's computer systems, which could then organize their own proportionate response. IOT can work together with voice alarm system to inform occupants the best escape routes, based on the direction of the fire spread.

8.2.2.2 Fire suppression

IOT technology can link a fire alarm or carbon monoxide detector with home appliances. If the system detects the presence of fire or carbon monoxide, it can automatically shut off these ignition sources. IOT can also work together with sprinkler systems with more targeted firefighting capabilities, helping to put out small fires and stem the tide until emergency crews arrive. By sensing exactly where the fire is, the nature of the fire, and whether there are any occupants in the room, a smart IOT-enabled fire system can choose different measures to specific rooms.

8.3 MACHINE LEARNING IN FIRE SAFETY DESIGN

ML is a part of AI technology which evolved from the study of pattern recognition and computational learning theory in artificial intelligence (AI). ML algorithms build a mathematical model based on sample data, known as "training data," which make predictions and decisions. It is closely related to computational statistics. It uses statistical techniques to let computers learn.

ML uses complex models and algorithms to predict. Therefore, the core of ML is to explore the study and construction of algorithms that can learn and make predictions on existing data. The study of mathematical optimization delivers methods, theories, and application domains to the field of ML. Data mining is another related field of study to ML.

These analytical models allow machine to decisions through learning from historical relationships and trends in the data. Therefore, ML has been used widely in almost all the industries recently, such as finance, security, and medical service.

In the past several years, ML has been widely used in many industries. Nowadays, in certain countries, ML has overtaken even experienced doctors in preliminary diagnosis of diseases. It is also widely used in finance (such as stock price prediction), cyber security (ML is used to predict the possible cyber-attacks), etc. Compared with a human being such as a data analyst or a doctor, the great advantage of AI lies in its capacity to process and explore extraordinarily large dataset and therefore, makes a much more accurate prediction than experienced professionals. The combination of the AI and Big Data supported by cloud computing further improves the capacity of AI. This is because AI can analyze a huge amount of Big Data only with the help of the large computational power provided by cloud computing. Therefore, the application of AI in the construction industry is one of the promising approaches for providing efficient and cost-effective design solutions. Using the machine to analyze large scope of scenarios and predict the accurate failure pattern of building under fire would be one of the promising solutions for performing an effective fire safety design of the building. The development of a new technique based on ML would enable a quicker assessment of a structure's vulnerability to fire and bring it within consideration for all steel-framed projects. Although construction research has considered ML for more than two decades, it had rarely been applied to fire safety design of buildings (Fu, 2018, 2020).

8.3.1 Machine learning and its application in the construction industry

However, as a traditional industry, AI and its applications in construction industry are far behind other areas. Although construction research has considered ML for more than two decades, it has rarely been applied to structural safety design. Some research has been undertaken in the past by using the ML for certain construction problems. Puri et al. (2018) used ML to predict the SPT N-value of soil using. Paudel et al. (2015) used ML for the prediction of building energy demand. Zhang et al. (2018) developed an ML framework for assessing post-earthquake structural safety. Shi et al. (2018) set up an evaluation model to assess the intelligent development of 151 cities in China using back-propagation neural network theory. Tixier et al. (2018) used Random Forest (RF) and Stochastic Gradient Tree Boosting (SGTB) methods to predict the injury in the construction sites. Özturan et al. (2018) used the artificial neural network to predict the concrete strength.

The primary problem that exists in the above application is that most of them can only solve simple problems such as prediction of the injury and relationships of the in-place density of soil or prediction of the strength of the concrete. In addition, few applications to fire safety design have been found. Therefore, further research on ML in solving more complex construction problems is imperative.

8.3.2 Problems experienced in the conventional structural fire analysis approach

As introduced in the preceding chapters, one of the primary fire safety design measures is to place fire protection on the structural members. As introduced in Chapter 5, in the project of the Shard, all primarily beams are protected for at least 60 min.

However, due to limitations in understanding the accurate response and failure patterns of the buildings, there is a possibility of overdesign due to addition of fire protection to unnecessary structural members. In certain cases, unnecessarily thick fire protection is used, which results in huge waste. Although an excess of money is spent due to overdesign of the structural fire protection, in certain circumstances, the building may not sufficiently be safe because of the wrong fire protection regime chosen. The above reported issues are primarily due to the limited capacity of an engineer to fully understand the failure pattern of a building in the fire conditions.

On the other hand, the present deterministic methods for assessing the fire protection regimes are time consuming. This is because fire development and subsequent structural response depend upon numerous factors. The appraisal of structural response in fire is challenging given the sources of uncertainty. In addition, for a structure such as a tall building, the structure systems are much more complicated, which also brings additional difficulties to the structural fire analysis.

The traditional design process is therefore time consuming and is limited by the ability of an engineer to fully understand the failure potential of the structure under different fire loadings. One of the best solutions to tackle above issues is to use ML.

8.3.3 Predicting failure patterns of simple steel-framed buildings in fire

Fu (2018, 2020) developed a new ML framework for fast prediction of the failure patterns of simple steel-framed buildings in fire and a potential assessment of subsequent progressive collapse. Critical temperature method introduced in Chapter 4 is used to define the failure patterns of each structural member. The training set of failure patterns is generated using both the Monte Carlo simulation and random sampling, which can guarantee a robust and sufficient large dataset for training and testing, thus guaranteying the accurate prediction. Three classifiers are chosen for prediction of failure patterns of buildings under fire: Decision Tree, *k*-nearest neighbor (*k*NN), and Neural Network using Google Keras with TensorFlow which is specially used for Google Brain Team. The ML framework is implemented using codes programmed by the author in VBA and Python language. A case study of a two-storey by two-bay steel-framed building was made. Two different fire scenarios were chosen. The procedure, shown in Figure 8.1,

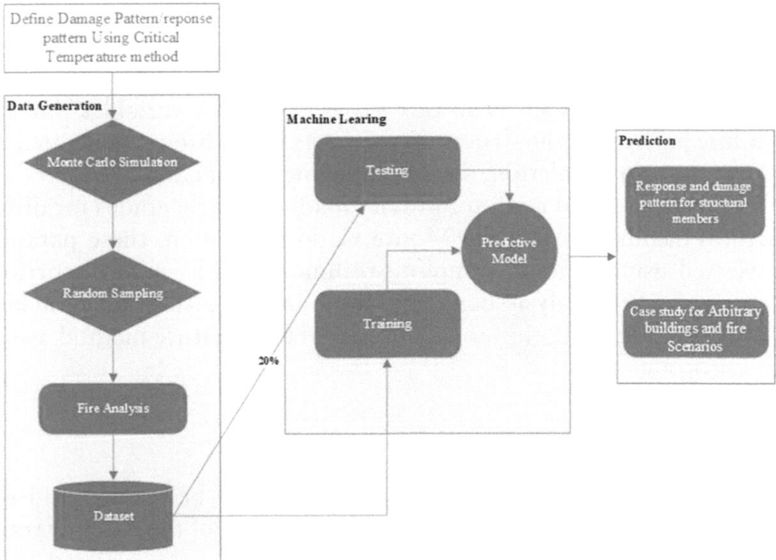

Figure 8.1 ML framework for the fire safety assessment of buildings (Fu, 2018).

gives satisfactory prediction of the failure pattern and collapse potential of the building under fire.

8.3.3.1 Define failure pattern

To enable ML, the failure patterns of the structure need to be first classified. There are different failure patterns when buildings are in fire. They have been introduced in the preceding chapters. Due to the complexity, it is difficult to digitalize and quantify these failure patterns to make the machine understand. One possible solution is to digitalize the failure patterns and subsequent image recognition, but it will be computationally difficult. However, in a fire safety design, the primary job for an engineer is to judge whether a member will fail. Therefore, a simplified critical temperature method introduced in Chapter 4 is used by Fu (2010, 2018) to define the failure patterns of each structural member.

8.3.3.2 Dataset generation using the Monte Carlo simulation and random sampling

To accurately predict the failure mode of one type of structure, the computer should be trained with a vast database of sufficient failure cases due to fire from real projects. The amount of the training cases is crucial for the accurate prediction result of the machine. However, there is no sufficient records of fire incidents available across the world; therefore, it will be hard to find sufficient training cases in the construction industry for the fire-induced failure modes.

To tackle this problem, Fu (2018, 2020) developed a novel method based on the Monte Carlo simulation, and random sampling is developed to generate sufficient large dataset in this project. The key variables that affect the failure patterns of the structural members under fire are generated using the Monte Carlo simulation, such as opening factors and fire load density (for fire load), imposed load (for gravity load), and steel grades for different structural members. After the Monte Carlo simulation, these parameters are selected using random sampling techniques with equal opportunities for structural fire analysis based on the Eurocode, and failure judgment can also be made using the so-called critical temperature method available from Eurocode.

8.3.3.3 Training and testing

After the dataset is generated, the computer shall be trained and tested using the dataset. Data scientists normally use 80% of the data for training and 20% of the data for testing.

8.3.3.4 *Failure pattern prediction*

If the testing results are satisfactory, the dataset generated is used for prediction. Different algorithms can be used for learning and prediction. Fu (2018, 2020) used *k*NN, Decision Trees, and Neural Network. Both Neural Network and *k*NN provide satisfactory prediction results.

8.3.3.5 *Fire safety design and progressive collapse potential check based on prediction results*

Based on the prediction of each single member, the collapse potential of the building can be checked using the member removal method specified by GSA (2003) and DoD (2009) the collapse potential can be assessed. Based on the prediction of the response of each structural member by the machine, a viable fire protection scheme can be selected.

8.3.4 Predicting and preventing fires with machine learning

Professor Jae Seung Lee and his students at Hongik University studied the information held by the city's fire service. Using ML, the university team was able to predict the probability of fires with an impressive 90% accuracy. The data supplied to the algorithm told students which neighborhoods were most at risk, allowing the Seoul Metropolitan Fire and Disaster Management Headquarters to deploy firefighters and patrol in the most vulnerable regions.

8.3.5 Machine learning of fire hazard model simulations for use in probabilistic safety assessments at nuclear power plants

Worrell et al. (2019) explored the application of ML to generate accurate and efficient metamodels for probabilistic fire safety assessments of nuclear power station. The process involved fire scenario definition, generating training data by iteratively running the fire hazard model called CFAST over a range of input spaces using the RAVEN software. The input and output data from a population of 675,000 CFAST were consolidated into a single comma-separated variable (.csv) file. R software was used for final metamodel selection and tuning.

They used both Decision Tree and *k*NN as prediction algorithms. It is found that a *k*NN model fit the vast majority of calculations within ±10% for maximum upper-layer temperature and its timing. The resulting *k*NN model was compared to an algebraic model typically used in fire probabilistic safety assessments. This comparison illustrated the potential of metamodels to improve modeling realism over simpler models selected for computational feasibility. While the *k*NN metamodel is a simplification of the higher-fidelity model, the error introduced is quantifiable and can be explicitly considered.

8.3.6 Learning algorithms and programming language

8.3.6.1 Learning algorithms

Up to date, three learning algorithms—Decision Tree, *k*NN, and Neural Network—are widely used for fire safety-related engineering problems. The research by Fu (2020) shows that *k*NN and Neural Network are the two promising classifiers for this particular engineering problem. Decision Tree yields less promising results. Worrell et al. (2019) also show the accuracy of *k*NN in their research.

8.3.6.2 Programming language

Python is one of the most popular languages for ML. It is available in various compliers such as Anaconda. It has most of the popular learning algorithms pre-programmed. Therefore, they are very handy for coding. Python libraries contain pre-written codes that can be imported into a code using Python's import feature. A Python framework is a collection of libraries intended to build a model of ML easily, without having to know the details of the underlying algorithms. An ML developer, should know how the algorithms work in order to know what results to expect, as well as how to validate them.

MATLAB is another widely used ML language package. Engineers and other domain experts have deployed thousands of ML applications using MATLAB.

R is an open-source language, so people can contribute from anywhere in the world. In R, the Black Box is referred to as a package. The package is nothing but a pre-written code that can be used repeatedly by anyone. There are many ML packages in R for a programmer to use.

REFERENCES

Fu, F. (2018), Fire safety assessment of buildings through machine learning, MSc thesis, University of Oxford.

Fu, F. (2020), Fire induced progressive collapse potential assessment of steel framed buildings using machine learning. *Journal of Constructional Steel Research*, 166, pp. 105918–105918.

GSA (2003), Progressive collapse analysis and design guidelines for new federal office buildings and major modernization projects. The U.S. General Services Administration.

ISO (1984), Paints and Varnishes—Colorimetry. Part 1: Principles. ISO 7724-1. ISO, Geneva.

Özturan, M., Kutlu, B., Özturan, T. (2008), Comparison of concrete strength prediction techniques with artificial neural network approach. *Building Research Journal*, 56, pp. 23–56.

Paudel, S., Nguyen, P. H., Kling Wil, L., Elmitri, M., Lacarri`ere, B., et al. (2015). Support Vector Machine in Prediction of Building Energy Demand Using Pseudo Dynamic Approach. *Proceedings of ECOS 2015 – The 28th International Conference on Efficiency, Cost, Optimization, Simulation and Environmental Impact of Energy Systems*, Jun 2015, Pau, France. hal-01178147.

Puri, N., Prasad, H. D., Jain, A. (2018), Prediction of geotechnical parameters using machine learning techniques. *Procedia Computer Science*, 125, pp. 509–517.

Shi, H. B., Tsai, S. B., Lin, X. W., Zhang, T. Y. (2018), How to evaluate smart cities' construction? A comparison of Chinese smart city evaluation methods based on PSF. *Sustainability*, 10(1), 37.

Unified Facilities Criteria, Department of Defense, UFC 4-023-03 (2003), Design of Buildings to Resist Progressive Collapse 14 July 2009 with Change 2n.

Worrell, C., Luangkesorn, L., Haight, J., Congedo, T. (2019, March), Machine learning of fire hazard model simulations for use in probabilistic safety assessments at nuclear power plants. *Reliability Engineering & System Safety*, 183, pp. 128–142.

Zhang, Y., Burton, H. V., Sun, H., Shokrabadi, M. (2018), A machine learning framework for assessing post-earthquake structural safety. *Structural Safety*, 72, pp. 1–16.

Chapter 9

Post-fire damage assessment

9.1 INTRODUCTION

Post-fire damage assessment is one of the key methods for retrofitting and reconstructing the structure after fire. Different damage assessment techniques including destructive and nondestructive assessment methods for concrete and steel structures will be introduced in this chapter.

9.2 POST-FIRE DAMAGE ASSESSMENT

Assessment of fire-damaged structures is essential because while the damages are often open to see, the structural damages that affect the elements and structures that provide support can't easily be seen.

Consequences of fire damage can include dramatically reduced strength of the steel reinforcement, prestressing or post-tensioning tendons; partial or total loss of the strength of concrete; and delamination of the concrete cover, which reduces its ability to protect the embedded steel.

There are several methods available for both concrete and steel structures, which will be introduced in this section.

9.2.1 Post-fire damage assessment of concrete structure

9.2.1.1 Visual inspection

The bearing capacity of the vertical structural elements (such as columns and shear walls) is critically important for the stability of a structure, and so they, as well as beams and slabs, should be visually inspected at the first instance.

9.2.1.1.1 Damage to columns

If the concrete cover had fallen from columns, the longitudinal reinforcement or stirrups should be checked for any damage.

9.2.1.1.2 Damage to shear walls

Spalling can be usually overserved in concrete walls, so it should be checked if there is any disengagement between the concrete and the reinforcement.

9.2.1.1.3 Damage to beams

Any decrease in the bearing capacity of beams will cause excessive deflections on the elements. In the event of fire, the bond between concrete and longitudinal reinforcement deteriorated, and hence the tension strength of longitudinal reinforcement decreases, which leads to large deflection of the beams. Therefore, the excessive deflection of the beams can be an indication of beam failure.

9.2.1.2 Schmidt rebound hammer

A Schmidt hammer can be used to determine estimated equivalent cube strengths of concrete members and assess the differences between fire-damaged and unaffected areas.

Schmidt hammer, shown in Figure 9.1, is a device used to measure the elastic properties and strength of concrete, mainly surface hardness and penetration resistance. The hammer measures the rebound of a spring-loaded mass impacting against the surface of a sample. The test hammer hits the concrete at a defined energy. Its rebound is dependent on the hardness of the concrete and is measured by the test equipment. By referring to a conversion chart, the rebound value can be used to determine the concrete's compressive strength.

9.2.1.3 Petrographic analysis

Petrographic analysis can determine the effects and extent of fire damage on a microscopic level. In petrographic analysis techniques, cast thin section, X-ray diffraction, and scanning electron microscope are the most widely used; electronic probe analysis is also sometimes used.

9.2.1.4 Spectrophotometer investigations

After fire, samples of concrete can be taken from the building for spectrophotometer testing. Color alternation in concrete is noticed, which can be used to judge temperature levels and the damage caused by the fire. The hue, value, and chroma of each sample can be measured by spectrophotometer according to ISO (1984). The results of this test can be used to find the depth of the concrete affected by the fire and, in turn, the thickness that should be repaired.

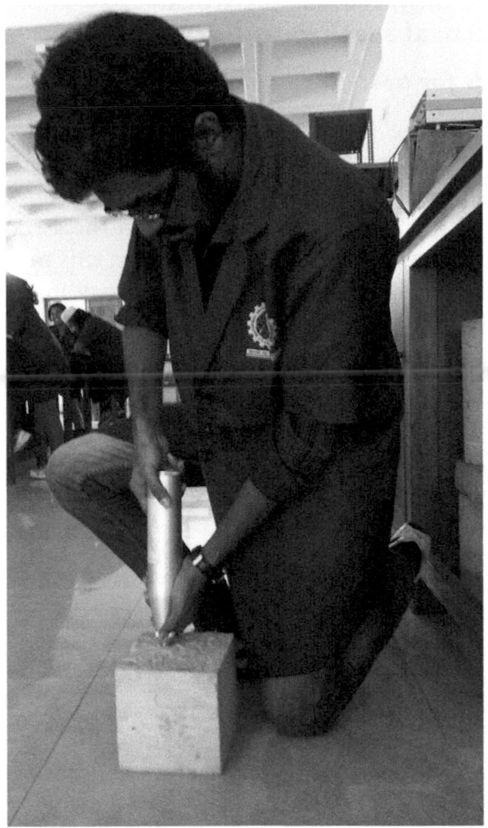

Figure 9.1 Schmidt rebound hammer. (This file is licensed under the Creative Commons Attribution-Share Alike 3.0 Unported license. https://upload.wikimedia.org/wikipedia/commons/1/1b/Schmidt_hammer_testing.jpg.)

9.2.1.5 Reinforcement sampling

Reinforcement samples are taken from the buildings after fire for yield strength and metallurgical testing. Thus, the residual strength of the reinforcement is determined.

9.2.1.6 Compression test

The strength values of the damaged concrete are obtained through direct compression test on the samples taken from the building exposed to fire. It is known that the compressive strength of the concrete not only degrades with temperature rise but also changes due to rate of heating, duration of the fire, loading, type of the aggregate, and water–cement ratio.

9.2.2 Post-fire damage assessment of structural steel members

9.2.2.1 Methods for post-fire damage assessment

The method for testing the residual strength of steel after fire includes on-site coupon tensile testing, chemical composition analysis method, and surface hardness method. The most accurate one is on-site coupon tensile testing. In this method, the coupon is cut from the structural members of the building to perform tensile testing. However, this method causes different degrees of damage to the structure, which may make post-fire restoration work difficult. The chemical composition analysis method also needs on-site sampling, and the process is tedious.

9.2.2.2 Nondestructive post-fire damage assessment of structural steel members using the Leeb harness method

For steel structure, surface hardness methods (including the Brinell hardness method, Rockwell hardness method, Victoria hardness method, and Leeb hardness method) are nondestructive, but only few of them have been used in post-fire damage assessment. Liu and Fu et al. (2020) developed a quick, simple, and efficient nondestructive detection method to measure the strength of steel after fire. It uses the so-called Leeb hardness method by means of establishing a relationship between the residual strength of steel members after fire and the Leeb hardness, and the post-fire steel strength can be fast determined without making any damage to the structural members.

As shown in Figure 9.2, the Leeb hardness testing is a nondestructive method for testing the strength of the steel members. It was invented by Dietamar Leeb in 1975. The method of the Leeb hardness testing is to drop certain weight of object through a tube to the surface of the specimens and test the impact velocity and velocity of the object at 1 mm distance from the surface when it bounces back.

A digital Leeb hardness tester TIME5351 is shown in Figure 9.2. The specimens were first grinded into smooth zones (30 mm×60 mm) for the test. The surface roughness was first assessed using roughness detectors. As shown in Figure 9.3, tests will be done for each smooth zone, and the average readings after removing the maximum and minimum values can be used.

A total of 120 Chinese H-shaped steel sections were selected for testing the Leeb hardness after fire by Liu and Fu (2020). Based on the test results, a linear regression analysis was made, and the correlations between the Leeb hardness and the residual tensile strength of flange and web are shown in Figures 9.4 and 9.5, respectively.

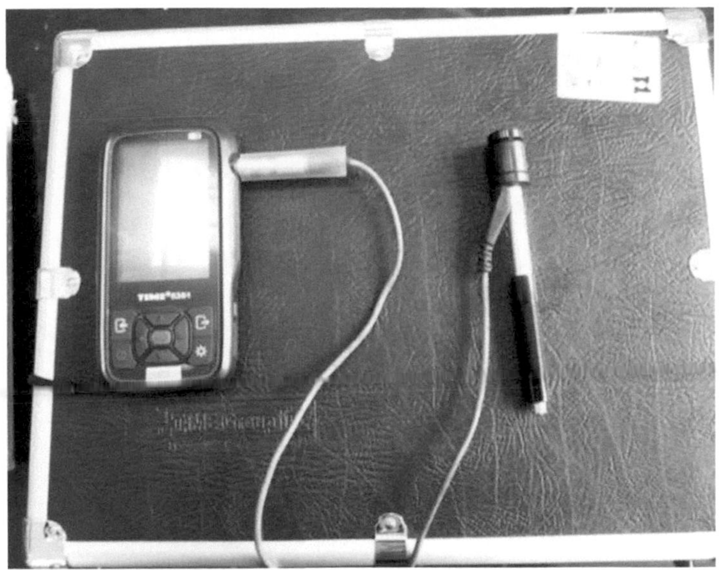

Figure 9.2 TIME5351 Leeb.

Figure 9.3 Leeb hardness testing of an I section.

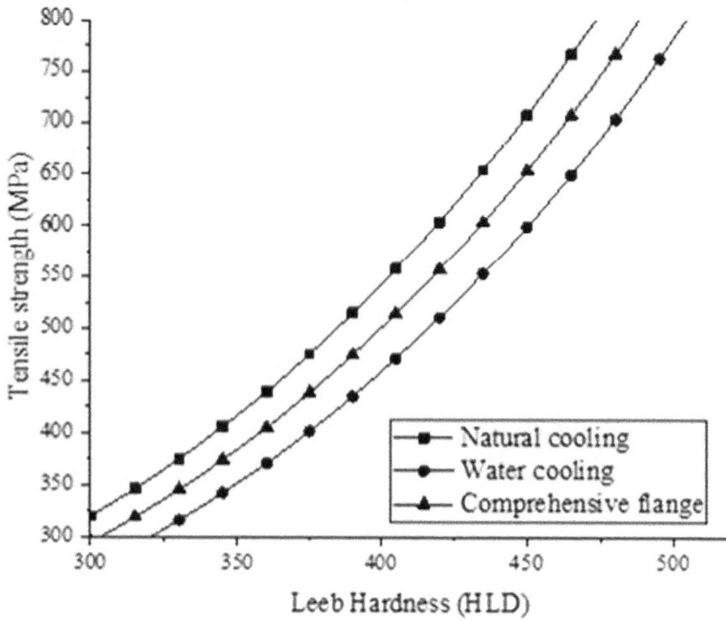

Figure 9.4 Correlation between the Leeb hardness and the residual tensile strength of flange Liu and Fu (2020).

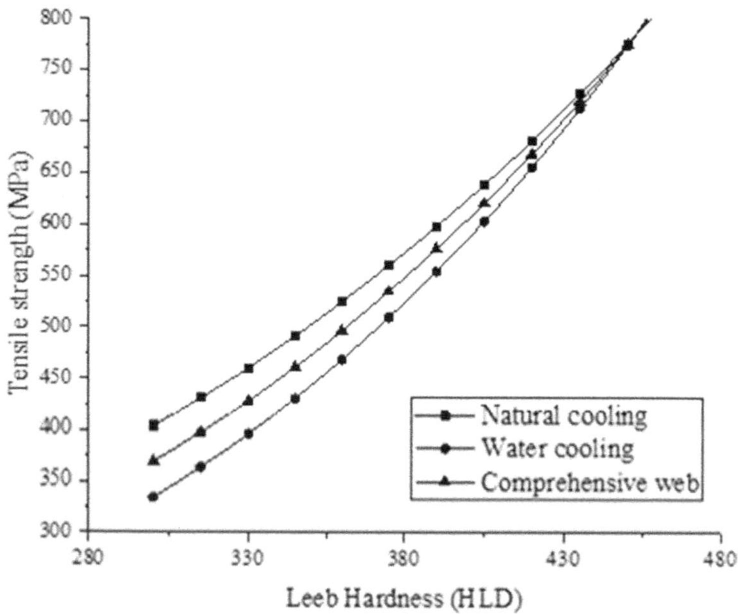

Figure 9.5 Correlation between the Leeb hardness and the residual tensile strength of web Liu and Fu (2020).

REFERENCES

Ada, M., Yüzer, N., Ayvaz, Y., Postfire damage assessment of a RC factory building. *Journal of Performance of Constructed Facilities*, 33(5), pp 04019047-1–04019047-12.

ISO (1984), Paints and varnishes—Colorimetry. Part 1: Principles. ISO 7724-1. ISO, Geneva.

Liu, D., Liu, X., Fu, F., Wang, W. (2020), Nondestructive post-fire damage assessment of structural steel members using Leeb harness method. *Fire Technology*, 56(4), pp. 1777–1799.

Index

T - #0129 - 111024 - C250 - 234/156/12 - PB - 9780367697716 - Gloss Lamination